Industrial Design Engineering

Inventive Problem Solving

Industrial Design Engineering
Inventive Problem Solving

John X. Wang

CRC Press
Taylor & Francis Group
Boca Raton London New York

CRC Press is an imprint of the
Taylor & Francis Group, an **Informa** business

CRC Press
Taylor & Francis Group
6000 Broken Sound Parkway NW, Suite 300
Boca Raton, FL 33487-2742

Library of Congress Cataloging-in-Publication Data

Names: Wang, John X., 1962- author.
Title: Industrial design engineering : inventive problem solving / John X. Wang.
Description: Boca Raton : CRC Press, 2017. | Includes index.
Identifiers: LCCN 2016035575| ISBN 9781498709590 (hardback : alk. paper) |
ISBN 9781498709606 (ebook)
Subjects: LCSH: Engineering design. | Industrial engineering. | Problem solving.
Classification: LCC TS171 .W36 2017 | DDC 620/.0042--dc23
LC record available at https://lccn.loc.gov/2016035575
No virus found in this message.
Checked by AVG - www.avg.com
Version: 2016.0.7797 / Virus Database: 4656/13094 - Release Date: 09/27/16

Visit the Taylor & Francis Web site at
http://www.taylorandfrancis.com

and the CRC Press Web site at
http://www.crcpress.com

Printed and bound in the United States of America by
Edwards Brothers Malloy on sustainably sourced paper

To the beautiful Sonny Wang Kindergarten
where engineering dreams start

Contents

Preface

Industrial engineering deals with the optimization of complex processes or systems. This engineering discipline has been driving the development, improvement, implementation, and evaluation of integrated systems. Forging the mathematical, physical, and social sciences together with the principles and methods of engineering design, industrial engineering has been applied to specify, predict, and evaluate the results to be obtained from such systems or processes. As an interdisciplinary engineering, its underlying concepts overlap considerably with business-oriented disciplines such as operations research and risk engineering.

Industrial design is a field with a large and extensive presence in our nation's manufacturing and services industries, as documented by the national datasets that provide the basis for this report. Designers are prolifically inventing new products, processes, and systems that have a profound impact on our economy and civil society. The National Endowment for the Arts (NEA) Design Program has been tracking numerous trends in the field of design, from the growing movement of design thinking to social impact design. Although this report brings together, for the first time, analytical perspectives regarding federal data on industrial design, it cannot be all-encompassing. This preface has benefited from conversations with some of the nation's leading designers, design curators, and design firms to convey information not captured by the report itself.

Industrial engineers are generally considered those professional individuals that develop the concepts for manufactured products such as cars, home appliances, and toys. How, nowadays, industrial designers often find themselves in a variety of roles and functions far beyond the development of manufactured products is summarized below:

- Rather than for manufacturers only, industrial engineers are working on projects for a variety of organizations, from government entities to private enterprises.
- The idea to utilize the design process as a way to analyze and innovate has dramatically changed the landscape of how industrial engineers work. For example, an industrial engineer might not

only design a radiation therapy machine for a hospital, but also the patient's interactive experience and touch points with medical staff in the emergency room.
- Similarly, industrial designers might work with retail merchandisers to reorganize store floor plans and re-imagine the in-store experience for potential customers.

Here, industrial designers are not just designing products, but designing user experiences, processes, and systems by applying the creative approach of what is described as "poetic design thinking," which enables industrial designers to work on diverse teams to solve these more complex challenges. This new book will present these new thinking paradigms systematically. Poetic design thinking is a tool to inspire innovation and influence systems change; industrial engineers are creative professionals who are doing just that.

John X. Wang
Featured Author and Poet
Grand Rapids, Michigan

Author

John X. Wang earned a PhD in reliability engineering from the University of Maryland, College Park, Maryland, in 1995. He was then with GE Transportation as an Engineering Six Sigma Black Belt, leading propulsion systems reliability and Design for Six Sigma (DFSS) projects while teaching GE-Gannon University's Graduate Co-Op programs and National Technological University professional short course, and serving as a member of the IEEE Reliability Society Risk Management Committee. Dr. Wang has worked as a Corporate Master Black Belt at Visteon Corporation, Reliability Engineering Manager at Whirlpool Corporation, E6 Reliability Engineer at Panduit Corp., and Principal Systems Engineer at Rockwell Collins. In 2009, he received an Individual Achievement Award when working as a Principal Systems Engineer at Raytheon Company. He joined GE Aviation Systems in 2010, where he was awarded the distinguished title of Principal Engineer— Reliability (CTH—Controlled Title Holder) in 2013.

As a Certified Reliability Engineer certified by the American Society for Quality, Dr. Wang has authored/co-authored numerous books and papers on reliability engineering, risk engineering, engineering decision making under uncertainty, robust design and Six Sigma, Lean manufacturing, and green electronics manufacturing. He has been affiliated with Austrian Aerospace Agency/European Space Agency, Vienna University of Technology, Swiss Federal Institute of Technology in Zurich, Paul Scherrer Institute in Switzerland, and Tsinghua University in China.

Having presented various professional short courses and seminars, Dr. Wang has performed joint research with the Delft University of Technology in the Netherlands and the Norwegian Institute of Technology.

Since his knowledge, expertise, and scientific results are well known internationally, Dr. Wang has been invited to present at various national and international engineering events.

As a highly accomplished inventor of various industrial designs and patent applications, Dr. Wang serves as an editor at Nano Research and Applications and is a member of BAOJ Nanotechnology Editorial Board.

Dr. Wang, a CRC Press Featured Author, has been a Top Contributor of LinkedIn's Poetry Editors & Poets group. Dr. Wang has contributed to the discussions including:

- Writing a sonnet is notoriously difficult due to the strict pentameter and rhyming pattern; does anyone prefer/enjoy writing this form of poetry?
- Do you proceed by images or by words when you write?

Connections

- CRC Press Featured Author: http://www.crcpress.com/authors/i351 -john-wang
- LinkedIn: http://www.linkedin.com/pub/john-wang/3/2b7/140
- Dr. John Wang's Amazon Author Central Profile: http://www.amazon .co.uk/-/e/B001H6WIPG

chapter one

Enduring sonnet
Evolving industrial design engineering

And from my pillow, looking forth by light

Of moon or favouring stars, I could behold

The antechapel where the statue stood

Of Newton with his prism and silent face,

The marble index of a mind for ever

Voyaging through strange seas of Thought, alone.

William Wordsworth
The Prelude (1850)

Industrial design engineers are continuing the voyage through strange seas of thought. The voyage is an enduring quest. The process of industrialization that began over 200 years ago is continuing to change the way people work and live, and doing it rapidly and globally. At the forefront of this movement is the profession of industrial engineering that develops and applies the technology that drives industrialization. This chapter describes how industrial design engineering evolved over the past two centuries developing methods and principles for the planning, design, and control of production and service systems. We will focus on the growth of the discipline that helped shape the industries worldwide and made substantial contributions to the industrialization of America and the world.

1.1 "A mind forever voyaging through strange seas of thought"

Logo design is an important area of industrial design engineering. Through logo design, a number of industrial designers have made such a significant impact on culture and daily life that their work is documented by historians of industrial science and engineering.

Raymond Loewy was a prolific American designer who is responsible for the Royal Dutch Shell corporate first logo. The original BP logo was in use until 2000. Apple's iconic logo has endured over the years. It's a slick graphic apple, almost perfect in symmetry save for its leaf and characteristic bite. Apple's enduring logo has evolved from Apple's first logo, which was greatly influenced by the "Enduring Sonnet" as an effective form for industrial communication (see Section 1.2).

Apple was founded as a partnership on April Fool's Day 1976 by three people who originally worked at Atari: Steven Gary Wozniak (1950–), Steven Paul Jobs (1955–2011), and Ronald Gerald Wayne (1934–). The first Apple logo was created in 1976 by former Atari draftsman/engineer Ron Wayne, who also wrote the Apple I manual and drafted the partnership agreement. Apple was incorporated on January 3, 1977, without Wayne, who sold his shares back.

In Apple, Ron Wayne's first order of business was designing a logo, and he created something he said was based on the personalities of both Mr. Jobs and Mr. Wozniak. Compared with Apple's iconic logo, a slick graphic apple, almost perfect in symmetry save for its leaf and characteristic bite, Apple's original logo had a completely different look—something you'd be more likely to find within the pages of a Victorian novel than adorning a piece of high-tech wizardry.

As shown in Figure 1.1, Apple's first logo was related to Newton and the falling apple. As a fan of poetry, Ron Wayne put it into a Gothic frame, and within that he took the last line from a Wordsworth sonnet—"A mind forever voyaging through strange seas of thought, alone."

Figure 1.1 Newton and the falling apple: The first Apple logo with the last line from a William Wordsworth sonnet.

1.2   Enduring sonnet: Most elastic form of industrial communication

Probabilistic poetic expression enables us to communicate the most serious of themes with the simplest language. Robert Frost's "Never Again Would Bird's Song Be the Same" exemplifies the "American sonnet" form.

As the most elastic form of communication, the sonnet can be traced back to the "Italian sonnet." Although there are earlier precedents, the first important sonneteers were Dante (1265–1321) and Francesco Petrarch (1304–1374). The Italian "sonnetto" maintains the following three elements:

- Octave: rhymed **abba abba**;
- Seset: rhymed more casually in any variation of *cde cde*; and
- Volta (or turn): the break between the two parts. Volta encourages a shift in tone.

The sonnet was brought to England through the translations of Petrarch by Wyatt and Surrey, written in the 1530s and 1540s and published in Tottel's "Miscellany" (1557). The "English sonnet," also known as the Shakespearean sonnet because of Shakespeare's mastery of the form, is composed of the following:

- Three quatrains: rhymed *abab, cdcd, efef*; and
- One terminal couplet: *gg*

Most of Shakespeare's sonnets were probably written in the 1590s. Sonnet 12 provides a classic example of Shakespearean construction (Figure 1.2).

Sonnet 12: When I do count the clock that tells the time (see Figure 1.2)

When I do count the clock that tells the time,
And see the brave day sunk in hideous night;
When I behold the violet past prime,
And sable curls ensilvered o'er with white;
When lofty trees I see barren of leaves,
Which erst from heat did canopy the herd,
And summer's green all girded up in sheaves
Borne on the bier with white and bristly beard:
Then of thy beauty do I question make
That thou among the wastes of time must go,
Since sweets and beauties do themselves forsake,
And die as fast as they see others grow;
And nothing 'gainst time's scythe can make defence

William Shakespeare (1564–1616)

Figure 1.2 When I do count the clock that tells the time.

Here, we may notice the following:

- Related to the theme of falling leaves with the west wind, we may notice that Sonnet 12 reflects on things' decay with time.
- The sonnet follows the pattern faithfully.
  - It has 14 regular end-stopped lines.
  - The first 8 lines are "soldered" together with "resonance."
  - Line 14 returns to the same "resonance" as line 8, answering the decay with "breed."
- The poem has a secondary "robust" structure, that of Italian sonnet (or Petrarchan sonnet), which gives it great internal strength.
  - Octave: lines 1–8 deal with decay in nature.
  - Seset: lines 9–14 deal with decay of mortal things.
  - Shakespeare honors the convention of the volta by adopting a "When... Then..." construction for the "action items" in the poem.
- Many variations on the motif "times' inevitable progress" throughout the poem:
  - Some refer to it as linear.
  - Some refer to it as circular.

Here, our poetic expression facilitates risk communication, an important part of industrial communication.

1.3   Evolution, engineering breakthrough, and industrial design engineering

The history of industrial design engineering is a history of knowledge. When the earliest civilizations appeared in Mesopotamia, Egypt, India, and China, they were largely constrained by their natural environment and by the climate. Philosophy, science, and art were largely determined by extra-human factors, such as seasons and floods. Over the course of many centuries, humans have managed to change the equation in their favor, reducing the impact of natural events on their civilization and increasing the impact of their civilization on nature.

This happened to be the history of industrial design engineering, an interdisciplinary field to become the "subject" of change, as opposed to being the "object" of change. The most important inventions date from prehistory. Here is a quick flashback:

- Tools, 2 million years ago, Africa
- Fire (heat generation), 1.9 million years ago, Africa
- Buildings 400,000 BC, France
- Burial, 70,000 BC, Germany
- Art, 28,000 BC
- Farming, 14,000 BC, Mesopotamia
- Animal domestication 12,000 BC
- Boat (8000 BC, Holland)
- Weapons (8000 BC); pottery, 7900 BC, China
- Weaving, 6500 BC, Palestine
- Money, sometime before the invention of writing, Mesopotamia
- Musical instruments, 5000 BC, Mesopotamia
- Metal and metal structure, 4500 BC, Egypt
- Wheel and transportation systems, 3500 BC, Mesopotamia
- Writing, 3300 BC, Mesopotamia (see Figure 1.3)
- Glass, 3000 BC, Phoenicia
- Sundial, 3000 BC, Egypt

During the Jemdet Nasr period (ca. 3100–2900 BC), writing was invented in Mesopotamia, perhaps in the city of Uruk (modern Warka), where the earliest inscribed clay tablets have been found in abundance. The Sumerian culture was not an isolated development but occurred during a period of profound transformations in politics, economy, and representational art. During the Uruk period of the fourth millennium BC, events included the following:

- The first Mesopotamian cities were settled.
- The first kings were crowned.
- A range of goods, from ceramic vessels to textiles, were mass-produced in state workshops.

Figure 1.3 Invention of writing, 3300 BC, Mesopotamia. (Cuneiform tablet: administrative account of barley distribution with cylinder seal impression of a male figure, hunting dogs, and boars; courtesy, the Metropolitan Museum of Art.) The clay tablet most likely documents grain distributed by a large temple, although the absence of verbs in early texts makes them difficult to interpret with certainty. The seal impression depicts a male figure guiding two dogs on a leash and hunting or herding boars in a marsh environment.

Early writing was used primarily as a means of recording and storing economic information. From the beginning a significant component of the written tradition consisted of lists of words and names that scribes needed to know in order to keep their accounts. Signs were drawn with a reed stylus on pillow-shaped tablets, most of which were only a few inches wide. The stylus left small marks in the clay which we call cuneiform, or wedge-shaped, writing.

1.4   The river: Where the first major civilizations were born

The first major civilizations were born in river valleys. Centralized authoritarian regimes are a direct consequence of large-scale irrigation agriculture: the problem of exploiting a river's power, that is, of building precise and timely waterworks, can only be solved by mass labor, by the mobilization and coordination of thousands of people, which is only possible in societies organized around centralized planning and capable of imposing absolute discipline. The bigger the river the greater the promise of wealth the stronger the "hydraulic state" has to be. The masses mobilized for waterworks can then be mobilized for other collective efforts, such as pyramids, temples, and fortifications. A navigable river then provided the infrastructure for interacting with other communities, that is, for both trade and warfare.

Once the infrastructure was in place, disciplines including industrial design engineering grew rapidly on all fronts:

- Agriculture
- Architecture from the ziggurat of the Sumerians to the pyramids of the Egyptians to the temples of the Greeks
- Bureaucracy from the city-states of the Sumerians to the kingdom of Egypt and the empire of Persia to the economic empire of Athens
- Politics from the theocracies of Mesopotamia and Egypt to the democracy of Athens
- Religion from the anthropomorphic deities of Mesopotamia to the complex metaphysics of Egypt, from the tolerant pantheon of the Greeks to the one God of the Persians and the Jews
- Writing and book authoring from the "Gilgamesh" in Mesopotamia to the "Adventures of Sinuhe" in Egypt to the "Bible" of the Jews to Homer's epics in Greece
- Economics from the agricultural societies of Mesopotamia and Egypt to the trade-based societies of Phoenicia and Athens
- Transportation systems from the horse-driven chariots of Mesopotamia to the Greek trireme
- Art from the funerary painting of the Egyptians to the realistic sculptures of the Greeks, etc.

As shown in Figure 1.4, the Pengtoushan culture was a Neolithic culture centered primarily on the central Yangtze River region in northwestern Hunan, China. It was roughly contemporaneous with its northern neighbor, the Peiligang culture. The two primary examples of Pengtoushan culture are the type site at Pengtoushan and the later site at Bashidang. The type site at Pengtoushan was discovered in Li County, Hunan. This site is the earliest permanently settled village yet discovered in China. Excavated in 1988, Pengtoushan has been difficult to date accurately, with a large variability in dates ranging from 9000 BC to 5500 BC. Cord-marked pottery was discovered among the burial goods. Analysis of Chinese rice residues which were Carbon-14 dated to 8200–7800 BC shows that rice had been domesticated by this time. The size of the Pengtoushan rice was larger than the size of naturally occurring wild rice; however, Pengtoushan lacked evidence of tools used in cultivating rice. Although not found at Pengtoushan, rice-cultivating tools were found in later sites associated with the Pengtoushan culture.

From ancient times, China displayed a holistic approach to nature, man, and government. Chinese religion realized the fundamental unity of the physical, the emotional and the social. Particularly during the Zhou dynasty, Chinese religion was natural philosophy. There was no

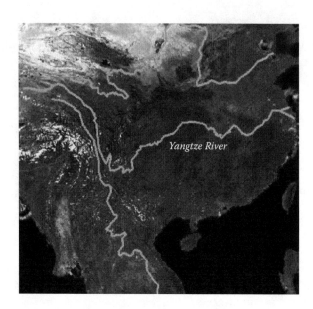

Figure 1.4 Yangtze River (China) where the Pengtoushan culture, a Neolithic culture, was born 7500–6100 BC. (Courtesy of U.S. Geological Survey [USGS], http://deltas.usgs.gov/rivers.aspx?river=yangtze.)

fear of damnation, no anxiety of salvation, no prophets, and no dogmas. Confucius was much more interested in the fate of society than in the fate of the souls of ordinary people. He believed that the power of example was the ideal foundation of the social contract: a ruler, a father, a husband has to "deserve" the obedience that is due to him. Thus, Confucius' philosophy was about the cultivation of the self, how to transform the ordinary individual into the ideal man. The ultimate goal of an individual's life is self-realization through socialization. If Confucius focused on society, Lao-tzu focused on nature. He believed in a "Tao," an ultimate unity that underlies the world's multiplicity. There is a fundamental reality in the continuous flow and change of the world: the "way" things do what they do. Understanding the "Tao" means identifying with the patterns of nature, achieving harmony with nature. The ideal course of action is "action through inaction" ("wuwei"): to flow with the natural order. The "Tao" is the infinite potential energy of the universe. "Qi" is vital energy/matter in constant flux that arises from the "Tao," and "Qi" is regulated by the opposites of "Yin" and "Yang." Everything is made of Yin and Yang.

1.5 *"La Mer": The ocean and civilization*

Debussy's "La Mer" is a unique mix of tone poem and symphony, a three-movement impression of the ocean. As the idea took shape in his mind, Debussy wrote to a friend in September 1903 that "I was destined for the fine career of a sailor," and that "only the accidents of life put me on another path." He acknowledged that a musical work about the ocean "could turn out to be like a studio landscape," but concluded that "I have countless reminiscences. This matters more, in my opinion, than a reality." The Debussy "La Mer" is a shimmering musical sketch inspired by the sea. The piece's ethereal impressionistic moods were groundbreaking, revealing the relationship between the ocean and civilization.

As shown by Figure 1.5, the Mediterranean Sea is a sea connected to the Atlantic Ocean surrounded by the Mediterranean region and almost completely enclosed by land: on the north by Anatolia and Europe, on the south by North Africa, and on the east by the Levant. The sea is technically a part of the Atlantic Ocean, although it is usually identified as a completely separate body of water. The name Mediterranean is derived from the Latin mediterraneus, meaning inland or in the middle of the earth (from "medius," middle, and "terra," earth). It was an important route for merchants and travelers of ancient times that allowed for trade and cultural exchange between emergent peoples of the region—the Mesopotamian, Egyptian, Phoenician, Carthaginian, Iberian, Greek, Macedonian, Illyrian, Thracian, Levantine, Gallic, Roman, Albanian, Armenian, Arabic, Berber, Jewish, Slavic, and Turkish cultures.

Maritime history dates back thousands of years. In ancient maritime history, evidence of maritime trade between civilizations dates back at

Figure 1.5 Carte nouvelle de la mer mediterranee Mediterranean Sea—map. (Courtesy of Library of Congress https://www.loc.gov/item/2001620436/.)

least two millennia. The first prehistoric boats are presumed to have been dugout canoes which were developed independently by various Stone Age populations. In ancient history, various vessels were used for coastal fishing and travel.

The Arabian Sea has been an important marine trade route since the era of the coastal sailing vessels from possibly as early as the third millennium BCE, certainly the late second millennium BCE through later days known as the Age of Sail. By the time of Julius Caesar, several well-established combined land-sea trade routes depended on water transport through the sea around the rough inland terrain features to its north. Navigation was known in Sumer between the fourth and the third millennium BCE, and was probably known by the Indians and the Chinese people before the Sumerians. The Egyptians had trade routes through the Red Sea, importing spices from the "Land of Punt," East Africa, and from Arabia.

1.6   Industrial design engineering and the advent of philosophy

A major step in the evolution of knowledge was the advent of philosophy. In Greece, Pythagoras was perhaps the first philosopher to speculate about the immortality of the soul. Heraclitus could not believe in the immortality of anything because he noticed that everything changes all the time:

- "You cannot enter the same river twice."
- "We are and we are not."

On the contrary, Parmenides, the most "Indian" of the Greek philosophers, believed that nothing ever changes: there is only one, infinite, eternal and indivisible reality, and we are part of this unchanging "one," despite the illusion of a changing world that comes from our senses. Zeno even proved the impossibility of change with his famous paradoxes.

For example, fast Achilles can never catch up with a slow turtle if the turtle starts ahead, because Achilles has to reach the current position of the turtle before passing it, and, when he does, the turtle has already moved ahead, a process that can be repeated forever. Democritus argued in favor of atomism and materialism: everything is made of atoms, including the soul. Socrates was a philosopher of wisdom, and noticed that wisdom knows what one does not know. His trial signaled the end of the dictatorship of traditional religion. Plato ruled out the senses as a reliable source of knowledge, and focused instead on "ideas," which exist in a world of their own, are eternal, and are unchangeable. He too believed in an immortal soul, trapped in a mortal body. By increasing its knowledge, the soul can become one with the ultimate idea of the universe, the idea of all ideas. On the contrary, Aristotle believed that knowledge "only" comes from the senses, and a mind is physically shaped by perceptions over a lifetime. He proceeded to create different disciplines to study different kinds of knowledge.

The Hellenistic age that followed Alexander's unification of the "oikoumene," the world that the Greeks knew, on a level never seen before fostered a new synthesis of views of the world. Hellenistic philosophy placed more emphasis on happiness of the individual, while Hellenistic religion placed more emphasis on salvation of the individual. Cynics, who thought that knowledge is impossible, saw attachment to material things as the root problem, and advocated a return to nature. Skeptics, who agreed that knowledge is impossible, thought that the search for knowledge causes angst, and therefore one should avoid having beliefs of any sort. Epicureans had a material view of the world: the universe is a machine and humans have no special status.

Epicureans claimed that superstitions and fear of death cause angst. Stoics viewed the entire universe as a manifestation of god and happiness as surrendering the self to the divine order of the cosmos, as living in harmony with nature.

From the very beginning, knowledge was also the by-product of the human quest for an answer to the fundamental questions:

- Why are we here?
- What is the meaning of our lives?
- What happens when we die?
- Is it possible that we live forever in some other form?

Humans used knowledge to reach different conclusions about these themes. The civilizations of Mesopotamia were mainly interested in "this" life. The Egyptians were obsessed with the afterlife, with immortality originally granted only to the pharaoh but eventually extended to everybody via the mysteries of Osiris, the first major ritual about the resurrection. The ancient Greeks did not care much for immortality, as Ulysses showed when he declined the goddess' invitation to spend eternity with her and preferred to return to his home; but later, in the Hellenistic period, a number of religious cults focused on resurrection: the Eleusinian mysteries about Demeter's search through the underworld for her daughter Persephone, the Orphic mysteries about Orpheus' attempt to bring back his wife Eurydice from the underworld, and the Dionysian mysteries about Dionysus, resurrected by his father Zeus. The Romans cared for the immortality of their empire, and were resigned to the mortality of the individual; but it was under Roman rule that a new Jewish religion,

Christianity, was founded on the notion that Jesus' death and resurrection can save all humans.

The other great theme of knowledge was the universe: what is the structure of the world that we live in? Neither the Indian nor the Greek philosophers could provide credible answers. They could only speculate. Nonetheless, the Hellenistic age fostered progress in mathematics such as Euclid's "Geometry" and Diophantus' "Arithmetic." They also nurtured science:

- Erarosthenes' calculation of the circumference of the Earth
- Archimedes' laws of mechanics and hydrostatics
- Aristarchus' heliocentric theory
- Ptolemy's geocentric theory

The Romans' main contribution to the history of knowledge may well be engineering, which is but the practical application of science to daily life. The Romans, ever the practical people, made a quantum leap in construction: from aqueducts to public baths, from villas to amphitheaters. At the same time, they too created a new level of unification: the unification of the Mediterranean world.

The intellectual orgy of Greek philosophy opened the Western mind. The Islamic philosophers felt the need to reconcile Islam and Greek philosophy. The two who exerted the strongest influence on the West, Abu Ali al-Husain ibn Abdallah ibn Sina Avicenna and Abu al-Walid Muhammad ibn Ahmad ibn Muhammad ibn Rushd Averroes, achieved such a momentous unification of religion and philosophy by envisioning the universe as a series of emanations from Allah, from the first intelligence to the intelligence of humans. This allowed them to claim that there is only one truth that appears like two truths: religion for the uneducated masses and philosophy for the educated elite. But there is no conflict between reason and revelation: ultimately, they both reach the same conclusions about the existence of Allah. The sufists, best represented by Ibn Arabi, added an almost Buddhist element: human consciousness is a mirror of the universal, eternal, infinite consciousness of Allah. Allah reveals himself to himself through human consciousness. The Sufi wants to achieve a state of participation in the act of self-revelation. The human condition is one of longing, of both joy (for having experienced the divine) and sorrow (for having lost the divine).

The invasions of the "barbaric" people of the east, the Arab invasion from the south, and the wars against the Persian Empire led to the decadence of Roman civilization and to the "dark age" that lasted a few centuries. The obliteration of culture was such that, eventually, Europe had to relearn its philosophy, science, and mathematics from the Arabs.

1.7   Industrial design engineering and supply chain risk engineering

The Arab invasion disrupted the economic and political unity of the Mediterranean Sea, and the rise of the Frankish kingdom, soon to be renamed "Holy Roman Empire," a mostly landlocked empire, caused a redesign of the main trade routes away from the sea. Venice alone remained a sea-trading power, and, de facto, the only economic link between Holy and Eastern Roman Empires. This "inland" trade eventually caused a "commercial" revolution. Trade fairs appeared in Champagne, the Flanders, and northern Germany, creating a new kind of wealth in those regions. The Italian communes became rich enough to be able to afford their own armies and thus become de facto independent and develop economies entirely based on trade. In northern Europe, a new kind of town was born, which did not rely on the Mediterranean Sea. Both in the north and in the south, a real bourgeois class was born. The medieval town was organized around the merchants, and then the artisans and the peasants.

As the horse became the main element in warfare, the landowner became the most powerful warrior. A new kind of nobility was created, land-owning nobility. The collapse of central authority in Western Europe led to feudalism, a system in which the nobility enjoyed ever greater power and freedom, a global "political" revolution. Thus, the "medieval synthesis":

- Church
- Cities
- Kings (clergy, bourgeoisie, nobility)

However, a fourth element was even more important for the history of knowledge. As Rome decayed, and Alexandria and Antioch fell to the Muslims, the capital of Christian civilization moved to Constantinople (Byzantium). Despite the Greek influence, this cosmopolitan city created great art but little or no philosophy or science. It was left to the monasteries of Western Europe to preserve the speculative traditions of the Greek world, except that they were mainly used to prove the Christian dogma. Monasticism was nonetheless crucial for the development of philosophy, music, and painting. The anarchy of the "dark age" helped monasteries become a sort of refuge for the intellectuals. As the choice of lay society came down to being a warrior or a peasant, being a monk became a more and more appealing alternative. Eventually, the erudite atmosphere of the monasteries inspired the creation of universities. And universities conferred degrees that allowed graduates to teach in any

Christian country, thus fueling an "educational" revolution. Johannes Scotus Erigena, Peter Abelard, Thomas Aquinas, Johannes Eckhart, John Duns Scotus (the "scholastics") were some of the beneficiaries. Western philosophy restarted with them. As their inquiries into the nature of the world became more and more "logical," their demands on philosophy became stricter. Eventually, Roger Bacon came to advocate that science be founded on logic and observation; and William Occam came to advocate the separation of logic and metaphysics, that is, of science and church.

1.8   Industrial design engineering and evolution technological revolution

1.8.1   Agricultural revolution

The commercial revolution of the new towns was matched by an agricultural revolution of the new manors. The plough, the first application of nonhuman power to agriculture, the four-field rotation (wheat, oats, legumes, and fallow), and the horseshoe caused an agricultural revolution in northern Europe that fostered rapid urbanization and higher standards of living. Improved agricultural techniques motivated the expansion of arable land via massive deforestation.

1.8.2   Technological revolution

In the cities, a technological revolution took place. It started with the technology of the mill, which was pioneered by the monasteries. Mills became pervasive for grinding grain, fulling clothes, pressing olives, and tanning. Textile manufacturing was improved by the spinning wheel, the first instance of belt transmission of power. And that was only the most popular instance of a machine because this was the first age of the machines. The mechanical clock was the first machine made entirely of metal.

1.8.3   Military revolution

There also was a military revolution, due to the arrival of gunpowder. Milan became the center of weapon and armor manufacturing. Demand for cannons and handguns created a whole new industry.

1.8.4   Engineering and artistic revolution

Finally, an "engineering/artistic" revolution also took place, as more and more daring cathedrals started dotting the landscape of Christianity. Each cathedral was an example of "total art," encompassing architecture, sculpture, painting, carpentry, and glasswork. The construction of a cathedral was a massive enterprise that involved masons, workers, quarrymen, smiths, carpenters, and so on. Not since the Egyptian pyramids had something so spectacular been tried. Each cathedral was a veritable summa of European civilization.

The political, commercial, agricultural, educational, technological, and artistic revolutions of the Middle Ages converged in the thirteenth century, the "golden century," to create an economic boom as it had not been seen for almost a millennium.

Improved communications between Europe and Asia, thanks to the Mongol Empire that had made travel safe from the Middle East to China, particularly on the "silk road," and to the decline of the Viking and Saracen pirates, led to a revival of sea trade, especially by the Italian city-states that profited from a triangular trade Byzantium-Arabs-Italy.

Florence, benefiting from the trade of wool, and Venice, benefiting from the sea trade with the East, became capitalistic empires. Venice sponsored technological innovation that enabled long-distance and winter voyages, while Florence sponsored financial innovation that enabled people to lend/borrow and invest capital worldwide. The Italian cities had a vested interest in improved education, as they needed people skilled in geography, writing, accounting, technology, and so on. It is not a coincidence that the first universities were established in Italy.

The economic boom came to an abrupt stop by plague epidemics, the "Black Death" that decimated the European population. But the Black Death also had its beneficial effects. The dramatic decrease in population led to a higher standard of living for the survivors, as the farmers obtained more land per capita and the city dwellers could command higher wages. The higher cost of labor prompted investments in technological innovation. At the same time, wealthy people bequeathed their fortunes to the creation of national universities which greatly increased the demand for books. The scarcity of educated people prompted the adoption of vernacular languages instead of Latin in the universities.

Throughout the Middle Ages, the national literatures had produced national epics such as

- "Beowulf" (900, Britain)
- "Edda" (1100, Scandinavia)
- "Cantar del Cid" (1140, Spain)
- Chretien de Troyes' "Perceval" (1175, France)
- "Slovo o Ploku Igoreve" (1185, Russia)
- "Nibelungen" (1205, Germany)
- "Chanson de Roland" (1200, France)
- Wolfram Von Eschenbach's "Parzival" (1210, Germany)
- Dante Alighieri's "Divine Comedy" (1300)

These national epics heralded a new age, in which the vernacular was used for the highest possible artistic aims, a veritable compendium of knowledge. After languishing for centuries, European poetry bloomed with

- Francesco Petrarca's "Canti" (1374, Italy)
- Geoffrey Chaucer's "Canterbury Tales" (1400, England)
- Inigo Santillana's "Cancionero" (1449, Spain)
- Francois de Villon's "Testament" (1462, France)
- Giovanni Boccaccio's "Decameron" (1353, Italy)

"Decameron" laid the foundations for narrative prose. In observance with the diktat of the Second Council of Nicaea (787), that the visual artist must work for the Church and remain faithful to the letter of the Bible, medieval art was permeated by aesthetics of "imitation." Christian art was almost a reversal of Greek art because the emphasis shifted from the body (mortal, whose movement is driven by emotions) to the soul (immortal, immune to emotions), from realism and movement to spirituality and immanence. Christian art rediscovered Egyptian and Middle-Eastern simplicity via Byzantine art. Nonetheless, centuries of illuminated manuscripts, mosaics, frescoes, and icons eventually led to the revolution in painting best represented by Giotto's "Scrovegni Chapel" (1305).

While Italian artists were refounding Greco-Roman art based on mathematical relationships and a sense of three-dimensional space, as in

- Paolo Uccello's "Battle of St. Romano" (1456),
- Masaccio's "Trinity" (1427), and
- Piero della Francesca's "Holy Conversation" (1474).

Northern European painters became masters of a "photographic" realism as in

- Jan Van Eyck's "The Virgin of the Chancellor Rolin" (1436), and
- "The Arnolfini Marriage" (1434).

Before Europe had time to recover from the Black Death, the unity of the Mediterranean was shattered again by the fall of Byzantium (1453) and the emergence of the Ottoman Empire (a Muslim empire) as a major European power.

However, Europe was coming out of the "dark age" with a new awareness of the world. Marco Polo had brought news of the Far East. Albertus Magnus did not hesitate to state that the Earth is a sphere. Nicolas Oresme figured out that the rotation of the Earth on an axis explains the daily motion of the universe.

In China, the Han and Tang dynasties had been characterized by the emergence of a class of officials-scholars and by a cultural boom. The Sung dynasty amplified those social and cultural innovations. The scholar-officials became the dominant class in Chinese society. The state was run like an autocratic meritocracy, but nonetheless a meritocracy. As education was encouraged by the state, China experienced a rapid increase in literacy, which led to a large urban literate class. The level of competence by the ruling class fostered technological and agrarian innovations that created the most advanced agriculture, industry, and trade in the world.

When Europe was just beginning to get out of its "dark age," China was the world's most populous, prosperous, and cultured nation in the world. The Mongol invasion, the Yuan dynasty, did not change the character of that society. Instead, the Yuan dynasty added an element of peace: "Pax tatarica" guaranteed by the invincible Mongol armies.

1.9 Year 1492 new front of industrial design engineering

Luckily for Christian Europe, in 1492 Spain opened a new front of knowledge: having freed itself of the last Arab kingdom, it sponsored the journey of Christopher Columbus to the West Indies, which turned out to be a new continent. That more or less accidental event marked the beginning of the "colonial" era, of "world trade," and of the Atlantic slave trade; and, in general, of a whole new mindset. Other factors were shaping the European mind:

- Gutenberg's printing press (1456), which made it possible to satisfy the growing demand for books;
- Martin Luther's Reformation (1517), which freed the northern regions from Catholic dogma;
- Copernicus' heliocentric theory (1530), which removed the Earth (and thus man) from the center of the universe; and the advent of the nation states (France, Austria, Spain, England, and, later, Prussia).

However, it was not the small European nations that ruled the world at the end of the Middle Ages. The largest empires (the "gunpowder empires") were located outside Europe. Gunpowder was only one reason for their success. They had also mastered the skills of administering a strong, centralized bureaucracy required to support an expensive military. In general, they dwarfed Europe at one basic dimension: knowledge. While Europe was just coming out of its "dark age," the gunpowder empires were at their cultural peak. The Ottoman Empire, whose capital Istanbul was the largest city in Europe, was a melting pot of races, languages, and religions. It was a sophisticated urban society, rich in universities and libraries, devoted to mathematics, medicine, and manufacturing. The Safavid Empire of Persia, which controlled the silk trade, was a homogeneous state of Muslim Persians. The Mughal Empire of India, an Islamic state in a Hindu country, was also a melting pot of races, languages, and religions. Ming China was perhaps the most technologically and culturally advanced of all countries.

The small European countries could hardly match the knowledge and power of these empires. And, still, a small country like Portugal or Holland ended up controlling a larger territory, stretching multiple continents, than any of those empires. A disunited Europe of small and poor states caught up in an endless loop of intestine wars, speaking different languages, technologically backward, that had to import science, philosophy, and technology from the Muslims, which had fewer people and resources than the Asian empires, managed to conquer the entire world with the only notable exception of Japan. Perhaps the problem was with the large-scale bureaucracies of those Asian empires, that, in the long term, became less and less competitive, more and more obscurantist. In some cases, their multiethnic nature caused centrifugal forces. Or perhaps Europe benefited from its own anarchy: continuous warfare created continuous competition and a perennial arms race. Perhaps the fact that no European power decisively defeated the others provided a motivation to improve what was missing in the more stable empires of the East. After all, the long-range armed sailing ships, which opened the doors to extra-European colonization, were the product of military build-up. Soon, world trade came to be based on sea transportation, which was controlled by Europeans. The printing press, which the gunpowder empires were slow to adopt or even banned, slowly changed the balance of knowledge. World trade was creating more demand for technological innovation (and science), while the printing press was spreading knowledge throughout the continent. And all of this was funded with the wealth generated by colonialism. While the Asian empires were busy enjoying their stability, the small European countries were fighting for supremacy, anywhere anytime; and, eventually, they even overthrew those much larger empires.

1.10   Creation of renaissance art and industrial design

Nowhere was the apparent oxymoron more intriguing than in Italy, a fragmented, war-torn peninsula that, nonetheless, became the cultural center of Europe. On a smaller scale, it was the same paradox: the tiny states of Italy and the Netherlands were superior in the arts to the powerful kingdoms of Spain, France, and England. In this case, though, the reason is to be found in the socio-economic transformation of the Middle Ages that had introduced a new social class: the wealthy bourgeoisie. This class was more interested in the arts than the courts (which were mainly interested in warfare). The main "customer" of the arts was still the Church, but private patronage of art became more and more common. This, in turn, led to elite of art collectors and critics. Aesthetics led to appreciation of genius: originality, individuality, creativity. Medieval art was imitation; Renaissance art was creation.

Perhaps the greatest invention of the Renaissance was the most basic of all from the point of view of knowledge: the self. The Egyptians and the Greeks did not have a truly unified view of the self, a unique way to refer to the "I" who is the protagonist of a life and, incidentally, is also a walking body. The Greeks used different terms (pneuma, logos, nous, and psyche) to refer to different aspects of the "I." The Middle Ages were the formative stage of the self, when the "soul" came to be identified with the thinking "I." The Renaissance simply exalted that great medieval invention, the "I," which had long been enslaved to religion. The "I" was now free to express and affirm it.

In a nutshell, the "Rinascimento" (Renaissance art) adapted classical antiquity to biblical themes. This was its fundamental contradiction: a Christian art based on pagan art. An art that was invented (by the Greeks) to please the pagan gods and (by the Romans) to exalt pagan emperors was translated into an art to pay tribute to the Christian dogma. The masterpieces in painting include:

- Leonardo da Vinci's "The Last Supper" (1497), and
- Michelangelo Buonarroti's "The Universal Judgment" (1541).

Architects such as Donato Bramante and Gianlorenzo Bernini dramatically altered the urban landscapes. However, there was also an obsession with ordering space, as manifested in

- Sandro Botticelli's "Allegory of Spring" (1478), and
- Raffaello Sanzio's "The School of Athens" (1511).

In the Netherlands, Hieronymous Bosch's "The Garden of Delights" (1504) was perhaps the most fantastic piece of art in centuries.

The Renaissance segued into the Baroque age, whose opulence really signified the triumph of European royalty and religion. Aesthetically speaking, the baroque was a restoration of order after the creative disorder of the Renaissance. The least predictable of the visual arts remained painting, with

- Pieter Bruegel's "Triumph of Death" (1562)
- Domenico El Greco's "Toledo" (1599)

- Pieter Rubens' "Debarquement de Marie de Medicis" (1625)
- Rembrandt's "Nightwatch" (1642)
- Jan Vermeer's "Malkunst" (1666)

Gian Lorenzo Bernini, born December 8, 1598, was a seventeenth-century artist and architect. As a leading figure in the development of the Italian baroque style, his best-known sculptures often combined white and colored marble with bronze and stucco. The Cornaro Chapel's Ecstasy of St. Teresa and the David are displayed at the Borghese Gallery. In addition, Bernini designed the magnificent baldachin, or canopy, inside St. Peter's Basilica.

Also in Italy, Giovanni Palestrina, Claudio Monteverdi (1567), and Girolamo Frescobaldi (1583) laid the foundations for classical music and the opera. The national literary scenes bloomed. Masterpieces of poetry included:

- Ludovico Ariosto's "Orlando Furioso" (1532)
- Luiz Vas de Camoes' "Os Lusiadas" (1572)
- Torquato Tasso's "Gerusalemme Liberata" (1575)
- Pierre de Ronsard's "Sonnets pour Helene" (1578)
- John Donne's "Holy Sonnets" (1615)
- John Milton's "Paradise Lost" (1667)

Even more characteristic of the era was theater:

- Gil Vicente's "Auto da Barca do Inferno" (1516)
- Christopher Marlowe's "Faust" (1592)
- William Shakespeare's "Hamlet" (1601) and "King Lear" (1605)
- Lope de Vega Carpio's "Fuente Ovejuna" (1614)
- Pedro Calderon's "El Gran Teatro del Mundo" (1633)
- Moliere's "Le Misanthrope" (1666) and Jean Baptiste Racine's "Phedre" (1677)

Francois Rabelais' "Gargantua et Pantagruel" (1552) and Miguel Cervantes' "Don Quijote" (1615) laid the foundations of the novel.

1.11   Science's impact on industrial design engineering

Progress in science was as revolutionary as progress in the arts. Famous scientists include:

- Tycho Brahe, who discovered a new star.
- Johannes Kepler, who discovered the laws of planetary motion.

- Francis Bacon, who advocated knowledge based on objective empirical observation and inductive reasoning.
- Galileo Galilei, who envisioned that linear uniform motion is the natural motion of all objects and that forces cause acceleration. This is the same for all falling objects, that is, the same force must cause objects to fall.

Suddenly, the universe did not look like the perfect, eternal, static order that humans had been used to for centuries. Instead, it looked as disordered, imperfect, and dynamic as the human world. New inventions included:

- Telescope (1608)
- Microscope (1590s)
- Pendulum clock (1657)
- Thermometer (1611)
- Barometer (1644)

Both the self and the world were now open again to philosophical investigation. Rene Descartes neatly separated matter and mind, two different substances, each governed by its set of laws (physical or mental). While the material world, including the body, is ultimately a machine, the soul is not: it cannot be "reduced" to the material world. His "dualism" was opposed by Thomas Hobbes' "materialism," according to which the soul is merely a feature of the body and human behavior is caused by physical laws.

Baruch Spinoza disagreed with both. He thought that only one substance exists: God. Nature is God ("pantheism"). The universe is God. This one substance is neither physical nor mental, and it is both. Things and souls are (finite) aspects (or "modes") of that one (infinite) substance. Immortality is becoming one with God/Nature, realizing the eternity of everything.

Gottfried Leibniz went in the other direction: only minds exist, and everything has a mind. Matter is made of minds ("panpsychism").

- Minds come in degrees, starting with matter whose minds are very simple, and
- Minds end with God whose mind is infinite.

The universe is the set of all finite minds (or "monads") that God has created. Their actions have been predetermined by God. Monads are "clocks that strike hours together."

Clearly, the scientific study of reality depended on perception, on the reliability of the senses. John Locke thought that all knowledge derives from experience ("empiricism"), and noticed that we only know the ideas and sensations in our mind. Those ideas and sensations are produced by perceptions, but we will never know for sure what caused those perceptions, how reality truly is out there: we only know the ideas that are created in our mind. Ideas rule our minds.

On the contrary, George Berkeley started from the same premises:

All we know is our perceptions.

He reached the opposite conclusion:

Matter does not even exist, that only mind exists ("idealism").

Reality is inside our mind: an object is an experience. Objects do not exist apart from a subject that thinks them. The whole universe is a set of subjective experiences. Locke thought that we can never know how the world really is, but Berkeley replied that the world is exactly how it appears: it "is" what appears, and it is inside our mind. Our mind rules ideas.

David Hume increased the dose of skepticism: if all ideas come from perception, then mind is only a theater in which perceptions play their parts in rapid succession. The self is an illusion. Mental life is a series of thoughts, feelings, and sensations. A mind is a series of mental events. The mental events do exist. The self that is supposed to be thinking or feeling those mental events is a fiction.

Observation led physicists to their own view of the world. By studying gases, Robert Boyle concluded that matter must be made of innumerable elementary particles, or atoms. The features of an object are due to the features and to the motion of the particles that compose it.

Following Galileo's intuitions and adopting Boyle's atomistic view, Isaac Newton worked out a mathematical description of the motion of bodies in space and over time. He posited an absolute time and an absolute space, made of ordered instants and points. He assumed that forces can act at a distance, and introduced an invisible "gravitational force" as the cause of planetary motion. He thus unified terrestrial and celestial mechanics: all acceleration is caused by forces, the force that causes free fall being the gravitational force, that force being also the same force that causes the Earth to revolve around the Sun. Forces act on masses, a mass being the quantitative property that expressed Galileo's inertia (the property of a material object to either remain at rest or in a uniform motion in the absence of external forces). Philosophers had been speculating that the universe might be a machine, but Newton did not just speculate: he wrote down the formulas.

Significant innovations were also introduced, for the first time in a long time, in mathematics. Blaise Pascal invented the mathematical theory of probability (and built the first mechanical adding machine). Leibniz envisioned a universal language of logic (a "lingua characteristica") that would allow to derive all possible knowledge simply by applying combinatorial rules of logic. Arabic numbers had been adopted in the sixteenth century. Signs for addition, subtraction, and multiplication were introduced by Francois Vieta. John Napier invented logarithms. Descartes had developed analytical geometry, and Newton and Leibnitz independently developed calculus.

It might not be a coincidence that a similar scientific, mathematical approach can be found in the great composers of the era:

- Antonio Vivaldi
- George Frideric Handel
- Johann Sebastian Bach

The next big quantum leap in knowledge came with the "industrial" revolution. It is hard to pinpoint the birth date of the industrial revolution (in 1721 Thomas Lombe built perhaps the first factory in the world, in 1741 Lewis Paul opened the first cotton mill, in 1757 James Watt improved the steam engine), but it is clear where it happened: Manchester, England. That city benefited from a fortunate combination of factors: water mills, coal mines, Liverpool's port and, last but not least, clock-making technology (the earliest factory mechanics were clock-makers). These factors were all in the hands of the middle class, so it is not surprising that the middle class (not the aristocracy or the government) ended up managing most of the enterprises.

chapter two

Monte Carlo simulation

Would an industrial engineer flip a coin like a poet?

The end of our foundation is the knowledge of Causes, and secret motions of things; and the enlarging bounds of the Human Empire, to the effecting of all things possible.

Francis Bacon
New Atlantis (1627)

The Industrial Revolution was marked by the introduction of power-driven machinery and ushered in a time of strong economic development. It generally applied to the social and economic changes that marked the transition from a stable agricultural and commercial society to a modern industrial society relying on complex machinery rather than tools. It is used primarily to refer to the period in British history from the middle of the eighteenth century to the middle of the nineteenth century. The principle of the division of labor and the resulting specialization of skills can be found in many human activities, and there are records of its application to manufacturing in ancient Greece. The first unmistakable examples of manufacturing operations carefully designed to reduce production costs by specialized labor and the use of machines appeared in the eighteenth century in England. The rapid industrial growth that began in England during the middle of the eighteenth century then spread over the next 50 years to many other countries, including the United States.

2.1 Industrial Revolution and industrial design engineering

The Industrial Revolution also saw the development of the steam engine, which was an engine that used steam to perform work, a milestone for industrial design engineering. Steam engines were used in transportation and to power factories. As shown in Figure 2.1, the revolution depended on devices such as James Watt's steam engine, which were invented at a rapidly increasing rate during the period. Motion, or driving power, is

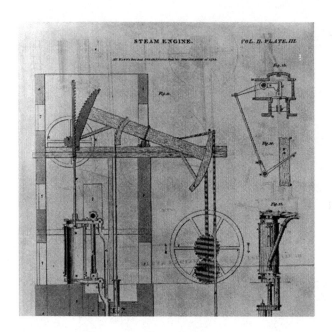

Figure 2.1 James Watt's steam engine. (Courtesy of Library of Congress, https://www.loc.gov/item/2006691752/.)

taken off from the moving engine by way of a belt attached to the crankshaft. This is opposed to a chain as was the case in the vertical steam engine that Thomas Newcomen constructed in the early eighteenth century to pump water from deep mine shafts. It was that engine that James Watt had improved upon over Newcomen's earlier design. After making major improvements in steam engine design in 1765, Watt continued his development and refinement of the engine until, in 1785, he successfully used one in a cotton mill. Once human, animal, and water power could be replaced with a reliable, low-cost source of motive energy, the Industrial Revolution was clearly established, and the next 200 years would witness invention and innovation the likes of which could never have been imagined.

The Industrial Revolution brought on a rapid concentration of people in cities and changed the nature of work for many people. The Industrial Revolution also saw the development of the steam engine, which was an engine that used steam to perform work. Steam engines were used in transportation and to power factories. In 1776 Adam Smith, in his *The Wealth of Nations*, observed the benefits of the specialization of labor in the manufacture of pins. Although earlier observers had noted this phenomenon, Smith's writings commanded widespread attention and helped foster an awareness of industrial production and broaden its appeal.

The eighteenth and nineteenth centuries brought much advancement to Britain and America. It was during this time period that the British Agricultural Revolution took place, which was a period of significant agricultural development marked by new farming techniques and inventions that led to a massive increase in food production. This agricultural growth created a ripple effect that spread throughout the countries. People were now able to leave the farms and move into cities because there was sufficient agricultural production to support life away from the farm. By the middle of the nineteenth century, the general concepts of division of labor, machine-assisted manufacture, and assembly of standardized parts were well established. Large factories were in operation on both sides of the Atlantic, and some industries, such as textiles and steel, were using processes, machinery, and equipment that would be recognizable even in the late twentieth century.

The conveniences of city life created a demand for other products, such as clothing. These products improved the quality of life. New technologies were invented to meet the growing demand for these products, which lead to the first industrial factories. Soon, people were moving into cities in greater numbers to find employment as factory workers. This ushered in the next great revolution, known as the Industrial Revolution. In this lesson, you will take a look at how industrialization and mass production furthered progress, and how they have impacted our environment.

The Industrial Revolution began in the late eighteenth and nineteenth centuries and was a period of significant economic development marked by the introduction of power-driven machinery. Much like the Agricultural Revolution, the Industrial Revolution began to take shape in Britain and then spread to other countries. During the Industrial Revolution, many power-driven machines were invented, which replaced hand tools. This included the cotton gin, which was a machine used to separate cotton fibers and their seeds. The American inventor of the cotton gin was a man by the name of Eli Whitney. The cotton gin, along with new inventions in spinning and weaving, made the mass production of cloth possible and gave a big boost to the textile industry. For mass production, industrial engineering was signalled by five important inventions in the textile industry:

1. John Kay's flying shuttle in 1733, which permitted the weaving of larger widths of cloth and significantly increased weaving speed;
2. Edmund Cartwright's power loom in 1785, which increased weaving speed still further;
3. James Hargreaves' spinning jenny in 1764;
4. Richard Arkwright's water frame in 1769; and
5. Samuel Crompton's spinning mule in 1779.

The last three inventions improved the speed and quality of thread-spinning operations. This era also led to a large increase in the use of coal. Coal replaced wood and other fuel sources because it was abundant, efficient, and required less work to mine than cutting wood. Coal was also used to make iron, which was used in the manufacturing of machines and tools, as well as the construction of ships and bridges.

2.2 Pioneering mass production methods for industrial design engineering

During the same period, similar ideas were being tried out in Europe. In England, Marc Brunel, a French-born inventor and engineer, established a production line to manufacture blocks (pulleys) for sailing ships, using the principles of division of labor and standardized parts. Brunel's machine tools were designed and built by Henry Maudslay, who has been called the father of the machine tool industry. Maudslay recognized the importance of precision tools that could produce identical parts; he and his student, Joseph Whitworth, also manufactured interchangeable, standardized metal bolts and nuts. The growth of manufacturing was accelerated by the rapid expansion of rail, barge, ship, and road transportation. The new transport companies not only enabled factories to obtain raw materials and to ship finished products over increasingly large distances, but they also created a substantial demand for the output of the new industries.

At this point in the Industrial Revolution, the methods and procedures used to organize human labor, to plan and control the flow of work, and to handle the myriad details on the shop floor were largely informal and were based on historical patterns and precedents. One man, Frederick W. Taylor, changed all of that.

In 1881, as shown in Figure 2.2, at the Midvale Steel Company in the United States, Frederick W. Taylor began studies of the organization of manufacturing operations that subsequently formed the foundation of modern production planning. After carefully studying the smallest parts of simple tasks, such as the shoveling of dry materials, Taylor was able to design methods and tools that permitted workers to produce significantly more with less physical effort. Later, by making detailed stopwatch measurements of the time required to perform each step of manufacture, Taylor brought a quantitative approach to the organization of production functions.

At the same time, Frank B. Gilbreth and his wife, Lillian M. Gilbreth, U.S. industrial engineers, began their pioneering studies of the movements by which people carry out tasks. Using the then-new technology of motion pictures, the Gilbreths analyzed the design of motion patterns and work areas with a view to achieving maximum economy of effort. The

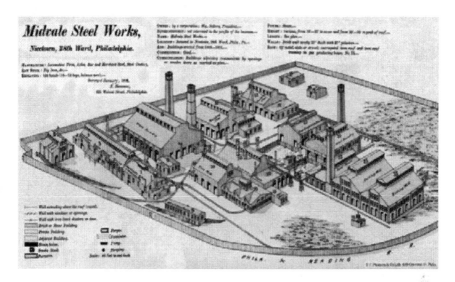

Figure 2.2 Midvale Steel Works aerial view, 1879 (Hexamer—Derived from map of Midvale Steel Works in Hexamer General Surveys, Volume 14, 1879).

"time-and-motion" studies of Taylor and the Gilbreths provided important tools for the design of contemporary manufacturing systems.

In 1916 Henri Fayol, who for many years had managed a large coal mining company in France, began publishing his ideas about the organization and supervision of work, and by 1925 he had enunciated several principles and functions of management. His idea of unity of command stated that an employee should receive orders from only one supervisor. This helped to clarify the organizational structure of many manufacturing operations.

2.3 Pioneering manufacturing engineering for industrial design engineering

Much of the credit for bringing these early concepts together in a coherent form, and creating the modern, integrated, mass production operation, belongs to the U.S. industrialist Henry Ford and his colleagues at the Ford Motor Company, where in 1913 a moving-belt conveyor was used in the assembly of flywheel magnetos. With it, assembly time was cut from 18 minutes per magneto to 5 minutes. The approach was then applied to automobile body and motor assembly. The design of these production lines was highly analytical and sought the optimum division of tasks among work stations, optimum line speed, optimum work height, and careful synchronization of simultaneous operations.

Born on July 30, 1863, on his family's farm in Dearborn, Michigan, Henry Ford enjoyed tinkering with machines from the time he was a young boy. Farm work and a job in a Detroit machine shop afforded him ample opportunities to experiment. He worked successively as an apprentice machinist, a part-time employee for the Westinghouse Engine Company, and an engineer with the Edison Illuminating Company. By then, he was earning enough money to experiment on building an internal combustion engine. By 1896, Ford had constructed his first horseless carriage, but he wanted to do even more!

"I will build a car for the great multitude," Ford proclaimed. At first, the automobile had been a luxury item only for the wealthy. Henry Ford wanted to create a car that ordinary people could afford, and in October 1908, he did it. The Model T sold for $850. In 19 years of manufacturing, Ford lowered the price to $260 and sold 15 million cars in the United States alone. How did he make the Model T so inexpensive?

As shown in Figure 2.3, Ford invented the modern assembly line. He doubled his workers' wages and cut the workday from 9 to 8 hours. Ford did this to ensure quality work and allow a three-shift workday. As a result, the company was able to make Model Ts 24 hours a day!

The success of Ford's operation led to the adoption of mass production principles by industry in the United States and Europe. The methods made major contributions to the large growth in manufacturing productivity

Figure 2.3 The first moving assembly line at Henry Ford's automobile factory in Detroit, Michigan. (Courtesy Library of Congress, http://www.americaslibrary .gov/es/mi/es_mi_detroit_1_e.html.)

that has characterized the twentieth century and produced phenomenal increases in material wealth and improvements in living standards in industrialized countries.

The automobile altered American society forever, changing where and how we lived. As more Americans owned cars, the organization of cities changed. The United States saw the growth of the suburbs and the creation of a national highway system. Americans were thrilled with the possibility of going anywhere, anytime. Ford witnessed many of these changes during his lifetime. In his later years, he spent most of his time working on Greenfield Village, a restored rural town modeled after his memories of Dearborn during his youth. Next time you are out on the road, try to imagine life without cars. Ask your family and friends how different they think the world would be.

2.4   Industrial Revolution and impacts on the environment and global economy

Even though this was a time of economic growth and development, the Industrial Revolution impacted the environment in negative ways. With the drudgery and toil of daily life made easier thanks to technological advancements, the world saw a major increase in population. This in and of itself led to environmental changes simply because there were more people consuming more natural resources. Not only was the population growing, but there was also a rapid growth in living standards thanks to the economic prosperity of this era. Higher living standards led to forests being cut down to make way for expanding cities and to provide lumber for construction. At this point in the Industrial Revolution, the methods and procedures used to organize human labor, to plan and control the flow of work, and to handle the myriad details on the shop floor were largely informal and were based on historical patterns and precedents.

Merlin Wilfred Donald (born November 17, 1939) is a Canadian psychologist and cognitive neuroscientist at Case Western University. Related to industrial design engineering, he noted the development of the human mind in the following four stages:

1. "Episodic" mind, which is limited to stimulus-response associations and cannot retrieve memories without environmental cues (lives entirely in the present).
2. "Mimetic" mind, capable of motor-based representations and of retrieving memories independently of environmental cues (understands the world, communicates and makes tools).
3. "Mythic" mind, which constructs narratives and creates myths.
4. "Theoretical" mind, capable of manipulating symbols.

One of the most tangible side-effects of the Industrial Revolution was the British Empire. There had been "empires" before and even larger ones (the Mongol empire). But never before had an empire stretched over so many continents:

- Africa
- America
- Oceania
- Asia

The Roman Empire had viewed itself as an exporter of "civilization" to the barbaric world, but the British Empire upped the ante by conceiving its imperialism as a self-appointed mission to redeem the world. Its empire was a fantastic business venture that exported people, capital, and goods, and created "world trade," not just regional trade. This enterprise was supported by a military might that was largely due to financial responsibility at home. Despite the fact that France had a larger population and more resources, Britain managed to defeat France in the War of the Spanish Succession (1702–1713), in the Seven Years' War (1756–1763), and in the Napoleonic wars (1795–1815).

Managing the British Empire was no easy task. One area that had to be vastly improved to manage a global empire was the area of global communications: steamships, railroads, the telegraph, the first undersea cable, and a national post system unified the colonies as one nation. They created the first worldwide logistical system. Coal, a key element in a country in which wood was scarce, generated additional momentum for the improvement of shipbuilding technology and the invention of railroads (1825).

Other areas that the British Empire needed to standardize were finance and law. Thus, the first global economic and legal systems were born. British economic supremacy lasted until 1869, when the first transcontinental railroad connecting the American prairies with the Atlantic Coast introduced a new formidable competitor: the United States.

No wonder, thus, that Adam Smith felt a new discipline had to be created, one that studied the dynamics of a complex economy based on the production and distribution of wealth. He explained the benefits of free competition and free trade, and how competition can work for the common good (as an "invisible hand").

2.5   Industrial Revolution and philosophy

Jeremy Bentham (1789) introduced "utilitarian" criteria to decide what is good and what is bad: goodness is what guarantees "the greatest happiness for the greatest number of people." The philosophy of "utilitarianism" was later perfected by John Stuart Mill, who wrote that "pleasure and freedom from pain are the only things desirable as ends," thus implying that good is whatever promotes pleasure and prevents pain.

France was much slower in adopting the Industrial Revolution, and never came even close to matching the pace of Britain industrialization, but the kingdom of the Bourbons went through a parallel "intellectual" revolution that was no less radical and influential: "Les Lumieres" (the lights), or the Enlightenment. It started in the salons of the aristocracy, usually run by the ladies, and then it spread throughout French society. The "philosophes" (philosophers) believed, first and foremost, in the power of reason and in knowledge, as opposed to the religious and political dogmas. They hailed progress and scorned conservative attitudes. The mood changed dramatically, as these philosophers were able to openly say things that a century earlier would have been anathema. Scientific discoveries (Copernicus, Galileo, and Newton), the exploration of the world, and the printing press led to cultural relativism: there are no dogmas, and only facts and logic should determine opinions. So they questioned authority across the board.

Julien LaMettrie was the ultimate materialist: he thought the mind is nothing but a machine (a computer, basically) and thoughts are due to the physical processes of the brain. There is nothing special about a mind or a life. Humans are just like all other animals. Charles Bonnet speculated that the mind may not be able to influence the body, but might simply be a side-effect of the brain ("epiphenomenalism").

The American Revolution (1776) was, ultimately, a practical application of the Enlightenment, a feasibility study of the ideas of the Enlightenment. The French Revolution (1789–1794) was a consequence of the new political discourse, but also signaled an alliance between the rising bourgeoisie, the starving peasants, and the exploited workers. By the turn of the century, the Enlightenment had also fathered a series of utopian ideologies, from Charles Fourier's phalanxes to Claude Saint-Simon's proto-socialism to Pierre Proudhon's anarchy.

In marked contrast with the British and French philosophers, the Germans developed a more "spiritual" and less "materialistic" philosophy. The Germans were less interested in economy, society, and politics, and much more interested in explaining the universe and the human mind, what we are and what is the thing out there that we perceive.

Immanuel Kant single-handedly framed the problem for future generations of philosophers. Noticing that the mind cannot perceive reality as it is, he believed that phenomena exist only insofar as the mind turns perceptions into ideas. The empirical world that appears to us is only a representation that takes place inside our mind. Our mind builds that representation thanks to some a-priori knowledge in the form of categories (such as space and time). These categories allow us to organize the chaotic flow of perceptions into an ordered meaningful world. Knowledge consists in categorizing perceptions. In other words, Kant said that knowledge depends on the structure of the mind.

Other German philosophers envisioned an even more "idealistic" philosophy.

Johann Fichte thought the natural world is construed by an infinite self as a challenge to itself and as a field in which to operate. The Self needs the non-Self in order to be.

Peter Schelling believed in a fundamental underlying unity of nature, which led to view nature as God, and to deny the distinction between subject and object.

The spiritual theory of reality reached its apex with Georg-Wilhelm-Friedrich Hegel. He too believed in the unity of nature, that only the absolute (infinite pure mind) exists, and that everything else is an illusion. He proved it by noticing that every "thesis" has an "antithesis" that can be resolved at a higher level by a "synthesis," and each synthesis becomes, in turn, a thesis with its own antithesis, which is resolved at a higher level of synthesis, and so forth. This endless loop leads to higher and higher levels of abstraction. The limit of this process is the synthesis of all syntheses: Hegel's absolute. Reality is the "dialectical" unfolding of the absolute. Since we are part of the absolute as we develop our dialectical knowledge, it is, in a sense, the absolute that is trying to know itself. We suffer because we are alienated from the absolute instead of being united with it. Hegel applied the same "dialectical" method to history, believing that history is due to the conflict of nations, conflicts that are resolved on a higher plane of political order.

Arthur Schopenhauer (1819) opened a new dimension to the "idealistic" discourse by arguing that a human being is both a "knower" and a "willer." As knowers, humans experience the world "from without" (the "cognitive" view). As free-willing beings, humans are also provided with a "view from within" (the "conative" view). The knowing intellect can only scratch the surface of reality; the will is able to grasp its essence. Unfortunately, the will's constant urge for ever more knowledge and action causes human unhappiness: we are victims of our insatiable will. In Buddhist-like fashion, Schopenhauer reasoned that the will is the origin of human sufferings: the less one "wills," the less one suffers. Salvation can come through a "euthanasia of the will."

Ludwig Feuerbach inverted Hegel's relationship between the individual and the Absolute and saw religion as a way to project the human experience ("species being") into the concept of God.

Soren Kierkegaard (1846) saw philosophy and science as vain and pointless because the thinker can never be a detached, objective, external observer: the thinker is someone who exists and is part of what is observed. Existence is both the thinker's object and condition. He thought that philosophers and scientists missed the point. What truly matters is the pathos of existing, not the truth of logic. Logic is defined by necessity, but existence is dominated by possibility. Necessity is a feature of being, possibility is a feature of becoming. He focused on the fact that existence is possibility, possibility means choice, and choice causes angst. We are trapped in an "aut-aut," between the aesthetic being (whose life is paralyzed by multiple possibilities) and the ethic being (whose life is committed to one choice). The only way out of the impasse is faith in God.

Inventions and discoveries of this age include:

- Alessandro Volta's battery, a device that converts chemical energy into electricity, and
- John Dalton's theory that matter is made of atoms of differing weights.

By taking Newton to the letter, Pierre-Simon LaPlace argued that the future is fully determined: given the initial conditions, every future event in the universe can be calculated. The primacy of empirical science ("positivism") was championed by Auguste Comte, who described the evolution of human civilization as three stages, corresponding to three stages of the human mind:

- Theological stage: events are explained by gods and kings rule;
- Abstract stage: events are explained by philosophy, and democracy rules; and
- Scientific ("positive") stage: there is no absolute truth, but science provides generalizations that can be applied to the real world.

Hermann von Helmholtz offered a detailed picture of how perception works, one that emphasized how an unconscious process in the brain was responsible for turning sense data into thought and for mediating between perception and action.

2.6 Influence of liberal arts and politics on industrial design engineering

In mathematics, George Boole resuscitated Leibniz's program of a "lingua characteristica" by applying algebraic methods to a variety of fields. His idea was that the systematic use of symbols eliminated the ambiguities of natural language. George Boole (1815–1864) was an English mathematician and a founder of the algebraic tradition in logic. He worked as a schoolmaster in England and from 1849 until his death as professor of mathematics at Queen's University, Cork, Ireland. He revolutionized logic by applying methods from the then-emerging field of symbolic algebra to logic. Where traditional (Aristotelian) logic relied on cataloging the valid syllogisms of various simple forms, Boole's method provided general algorithms in an algebraic language which applied to an infinite variety of arguments of arbitrary complexity. These results appeared in two major works, *The Mathematical Analysis of Logic* (1847) and *The Laws of Thought* (1854).

A number of mathematicians realized that the traditional (Euclidean) geometry was not the only possible geometry. Joseph Fourier discovered that any periodic function can be decomposed into sine and cosine functions. Non-Euclidean geometries were developed by Carl-Friedrich Gauss, Nikolaj Lobachevsky (1826), Janos Bolyai (1829), and Georg Riemann (1854). The latter realized that the flat space of Euclidean geometry (the flat space used by Newton) was not necessarily the only possible kind of space: space could be curved, and he developed a geometry for curved space (in which even a straight line is curved, by definition). Each point of that space can be more or less curved, according to a "curvature tensor."

While human knowledge was expanding so rapidly, literature was entering the "romantic" age. The great poets of the age were

- William Blake and William Wordsworth in England
- Friedrich Hoelderlin and Johann-Wolfgang Goethe in Germany
- Giacomo Leopardi in Italy

With the exception of Carlo Goldoni's comedies in Italy, theater was dominated by German drama:

- Gotthold-Ephraim Lessing in Germany
- Friedrich von Schiller and Georg Buchner

The novel became a genre of equal standing with poetry and theater via

- Goethe's *Wilhelm Meister* (1796)
- Stendhal's *Le Rouge et Le Noir* (1830)
- Honore' de Balzac's *Le Pere Goriot* (1834)
- Emily Bronte's *Wuthering Heights* (1847)
- Herman Melville's *Moby Dick* (1851)
- Nikolaj Gogol's *Dead Souls* (1852)
- Gustave Flaubert's *Madame Bovary* (1857)
- Victor Hugo's *Les Miserables* (1862)

While painting was relatively uneventful compared with the previous age, despite the originality of works such as Francisco Goya's "Aquelarre"

(1821) and Jean-Francois Millet's "The Gleaners" (1851), this was the age of classical music that boasted the geniuses of Wolfgang Amadeus Mozart, Franz-Peter Schubert, and Ludwig Van Beethoven.

In the meantime, the world had become a European world. The partition of Africa (1885) had given Congo to Belgium, Mozambique and Angola to Portugal, Namibia and Tanzania to Germany, Somalia to Italy, Western Africa and Madagascar to France, and then Egypt, Sudan, Nigeria, Uganda, Kenya, South Africa, Zambia, Zimbabwe, and Botswana to Britain. Then there were the "settler societies" created by the European immigrants who displaced the natives: Canada, United States, Australia, and South Africa. In subject societies such as India's (and, de facto, China's), few Europeans ruled over huge masses of natives. The mixed-race societies of Latin America were actually the least "European." There were fewer and shorter Intra-European wars but many more wars of conquest elsewhere. Europeans controlled about 35% of the planet in 1800, 67% in 1878, and 84% in 1914.

Japan was the notable exception. It had been the least "friendly" to the European traders, and it became the first (and only) non-European civilization to "modernize" rapidly. In a sense, it became a "nation" in the European sense of the word. It was also the first non-European nation to defeat a European power (Russia). No wonder that the Japanese came to see themselves as the saviors of Asia: they were the only ones that had resisted European colonization.

To ordinary people, the age of wars among the European powers seemed to be only a distant memory. The world was becoming more homogeneous and less dangerous. One could travel from Cairo to Cape Town, from Lisbon to Beijing with minimal formalities. It was "globalization" on a scale never seen before and not seen again for a century. Such a sense of security had not been felt since the days of the Roman Empire, although, invisible to most, this was also the age of a delirious arms race that the world had never seen before.

No wonder that the European population increased dramatically at the end of the nineteenth century. In 30 years, Germany's population grew by 43%, Austria-Hungary's by 35%, and Britain's by 26%. A continuous flow of people immigrated to the Americas.

After the French revolution, nationalism became the main factor of war. Wars were no longer feuds between kings, they were conflicts between peoples. This also led to national aspirations by the European peoples who did not have a country yet: notably Italians and Germans, who were finally united in 1861 and 1871 (but also the Jews, who had to wait much longer for a homeland). Nationalism was fed by mass education (history, geography, literature), which included, more or less subtly, the exaltation of the national past.

2.7   The influence of poetic thinking
on industrial design engineering

France lived its "Belle Epoque" (the 40 years of peace between 1871 and 1914). It was the age in which cafes (i.e., the lower classes) replaced the salons (i.e., the higher classes) as the cultural centers. And this new kind of cultural center witnessed an unprecedented convergence of art and politics. Poetry turned toward "Decadentism" and "Symbolism," movements pioneered by

- Charles Baudelaire's "Les Fleurs du Mal" (1857)
- Isidore de Lautreamont's "Les Chants de Maldoror" (1868)
- Arthur Rimbaud's "Le Bateau Ivre" (1871)
- Paul Verlaine's "Romances sans Paroles" (1874)
- Stephane Mallarme's "L'apres-midi d'un Faune" (1876)

Painters developed "Impressionism," which peaked with Claude Monet, and then "Cubism," which peaked with Pablo Picasso, and, in between, original styles were pursued by Pierre Renoir, Georges Seurat, Henry Rousseau, Paul Gaugin, and Henri Matisse. France had most of the influential artistic movements of the time. In the rest of Europe, painting relied on great individualities: Vincent van Gogh in Holland, Edvard Munch in Norway, Gustav Klimt in Austria, and Marc Chagall in Russia. French writers founded "Dadaism" (1916) and "Surrealism" (1924), and an Italian in Paris founded "Futurism" (1909). They inherited the principle of the "Philosophes": question authority and defy conventions, negate aesthetic and moral values. At the same time, they reacted against the ideological values of the Enlightenment itself: Dadaism exalted irrationality, Surrealism was fascinated by dreams, and Futurism worshipped machines.

Berlin, in the meantime, had become not only the capital of a united Germany but also the capital of electricity. Germany's pace of industrialization had been frantic. Werner Von Siemens founded Siemens in 1847. In 1866 that company invented the first practical dynamo. In 1879, Siemens demonstrated the first electric railway and in 1881 it demonstrated the first electric tram system. In 1887, Emil Rathenau founded Siemens' main competitor, the Algemeine Elektrizitats Gesellschaft (AEG), specializing in electrical engineering, whereas Siemens specialized in communication and information. In 1890, AEG developed the alternating-current motor (invented in the United States by Nikola Tesla) and the generator, which allowed building the first power plants: alternating current made it easier to transmit electricity over long distances. In 1910, Berlin was the greatest center of electrical production in the world. Germany's industrial output had passed France's (in 1875) and Britain's (in 1900). Berlin was becoming a megalopolis, as its population grew from 1.9 million in 1890 to 3 million in 1910.

Electricity changed the daily lives of millions of people, mainly in the United States, because it enabled the advent of appliances, for example,

- Josephine Cochrane's dishwasher (1886)
- Willis Carrier's air conditioner (1902)
- General Electric's commercial refrigerator (1911)

Life in the office also changed dramatically. First (in 1868) Christopher Latham Sholes introduced a practical typewriter that changed the concept of corresponding, and then (in 1885) William Burroughs introduced an adding machine that changed the concept of accounting. Progress in transportation continued with

- Daimler and Maybach's motorcycle (1885)
- Karl Benz's gasoline-powered car (1886), Ferdinand von Zeppelin's rigid dirigible (1900)
- Wilbur and Orville Wright's airplane (1903)

However, more importantly, the United States introduced a new kind of transportation, not physical (of people) but virtual (of information). The age of communications was born with

- Samuel Morse's telegraph (1844)
- Alexander Bell's telephone (1876)
- Thomas Edison's phonograph (1877)
- Kodak's first consumer camera (1886)

Just like Louis Daguerre, a French artist and photographer, had invented the "daguerrotype" in 1839, but his invention had been improved mainly in the United States, so the Lumiere brothers, French inventors and pioneer manufacturers of photographic equipment who devised an early motion-picture camera and projector called the Cinématographe, invented cinema (in 1895) but the new invention soon became an American phenomenon. The most dramatic of these events was perhaps Italian inventor Guglielmo Marconi's transatlantic radio transmission of 1901, when the world seemed to shrink. "Creationist" views of the world had already been attacked in France by the "philosophes." In the Age of Progress, a new, much more scientific attack came from Britain.

Herbert Spencer attempted a synthesis of human knowledge that led him to posit the formation of order as a pervasive feature of the universe. Basically, the universe is "programmed" to evolve toward more and more complex states. In particular, living matter continuously evolves. The fittest forms of life survive. Human progress (wealth, power) results from a similar survival of more advanced individuals, organizations, societies, and cultures over their inferior competitors.

Charles Darwin explained how animals evolved: through the combination of two processes, one of variation (the fact that children are not identical to the parents, and are not identical to each other) and selection (the fact that only some of the children survive). The indirect consequence of these two processes is "adaptation," whereby species tend to evolve toward the configuration that can best cope with the environment. The "struggle for survival" became one of the fundamental laws of life. In a sense, Darwin had merely transferred Adam Smith's economics to biology. But he had also introduced an important new paradigm: "design without a designer." Nature can create amazingly complex and efficient organisms without any need for a "designer" (whether human or divine). Humans are used to the idea that someone designs, and then builds, an artifact. A solution to a problem requires some planning. But Darwin showed that Nature uses a different paradigm: it lets species evolve through the combined forces of variation and selection, and the result is a very efficient solution to the problem (survival). No design and no planning are necessary. It was more than a theory of evolution: it was a new way of thinking that was immediately applied to economics, sociology, history, and so on.

Ernst Haeckel argued that "ontogeny recapitulates phylogeny": the development of the body in the individual of a species (or ontogeny) summarizes the evolutionary development of that species (phylogeny).

Far less publicized, but no less dramatic, was the discovery of Gregor Mendel. He set out to explain why children do not inherit the average of the traits of their parents (e.g., a color in between the black eyes of the mother and the blue eyes of the father) but only the trait of one or the other (black or blue eyes). He came up with a simple but, again, revolutionary explanation: there are units of transmission of traits (which today we call "genes"), and one inherits not a mathematical combination of one's parents' traits but either one or the other unit. Mendel introduced an important distinction: the "genotype" (as the program that determines how an organism looks like) versus "the phenotype" (the way the organism looks like).

Similar progress was going on in the study of the human mind. Paul Broca studied brain lesions to understand the structure of the brain and how it determines human behavior and personality.

2.8 The rise of physics on industrial design engineering

The acceleration in physics had been dramatic since Newton's unification of terrestrial and celestial mechanics, but the age of steam and electrical power introduced the first strains on its foundations. In 1824, Sadi Carnot had worked out a preliminary science of heat, or thermodynamics, and in 1864 James Clerk Maxwell unified electricity and magnetism, thus founding electromagnetism. In 1887, Heinrich Herz discovered radio waves, and in 1895 Wilhelm-Conrad Roentgen discovered x rays. Newton's physics had not been designed for these phenomena.

In 1896, radioactivity was discovered and led physicists to believe that the atom was not indivisible, that it had its own structure. In 1900, Max Planck invented quantum theory by positing that energy can only be transmitted in discrete "quanta." In 1905, Albert Einstein published "The Special Theory of Relativity." In 1911, Ernest Rutherford showed how the atom is made of a nucleus and orbiting electrons.

Newton's physics viewed the world as a static and reversible system that undergoes no evolution, whose information is constant in time. Newton's physics was the science of being. But his physics was not very useful to understand the world of machines, a dynamic world of becoming. Thermodynamics describes an evolving world in which irreversible processes occur: something changes and can never be undone. Thermodynamics was the science of becoming. The science of being and the science of becoming describe dual aspects of nature. Thermodynamics was born to study gases: systems made of myriad small particles in frantic motion. Newton's physics would require a dynamic equation for each of them, which is just not feasible. Thermodynamics describes a macroscopic system by global properties (such as temperature, pressure, volume). Global properties are due to the motion of its particles (e.g., temperature is the average kinetic energy of the molecules of a system). They are fundamentally stochastic, which implies that the same macro-state can be realized by different micro-states (e.g., a gas can have the same temperature at different points in time even if its internal status is changing all the time). Sadi Carnot realized that a "perpetual-motion" machine was not possible: it is not possible to continuously convert energy from one form to another and back. The reason is that any transformation of energy has a "cost" that came to be called "entropy." That quantity became the real oddity of thermodynamics. Everything else was due to a view of a complex system stochastic (as opposed to Newton's deterministic view of a simple system), but entropy was a new concept, which embodied a fundamental feature of our universe: things decay, and some processes are not reversible. Heat flows spontaneously from hot to cold bodies, but the opposite never occurs. You can dissolve a lump of sugar in a cup of coffee, but, once it is dissolved, you can never bring it back. You may calculate the amount of sugar, its temperature and many other properties, but you cannot bring it back. Some happenings cannot be undone. The second law of thermodynamics states that the entropy (of an isolated system) can never decrease. It is a feature of this universe that natural processes generate "entropy." This translates into a formula that is not equality. Newton's physics was built on the equal sign (something equals something else). Thermodynamics introduced the first law of nature that was an inequality.

Ludwig Boltzmann interpreted entropy as a measure of disorder in a system. He offered a statistical definition of entropy: the entropy of a macrostate is the logarithm of the number of its microstates. Entropy measures the very fact that many different microscopic states of a system may result in the same macroscopic state. One can interpret this fact as "how disordered the system is." This interpretation combined with the second law of thermodynamics led to the fear of an "eternal doom": the universe must evolve in the direction of higher and higher entropy, thus toward the state of maximum entropy, which is absolute disorder, or the "heat death."

Maxwell's electromagnetism introduced another paradigm shift: the concept of field pioneered by Michael Faraday. Maxwell proved that electricity and magnetism, apparently related to different phenomena, are, in reality, the same phenomenon. Depending on circumstances, one can witness only the electrical or only the magnetic side of things, but they actually coexist all the time. The electric force is created by changes in the magnetic field. The magnetic force is created by changes in the electric field. The oddity was that the mathematical expression of these relations between electric and magnetic forces turned out to be "field equations," equations describing not the motion of particles but the behavior of fields. Fields are generated by waves that radiate through space. Gravitation was no longer the only example of action at distance: just like there was a "gravitational field" associated to any mass, so there turned out to exist an "electromagnetic" field to any electrical charge. Light itself was shown to be made up of electromagnetic waves.

Ernst Mach (February 18, 1838–February 19, 1916) was another influential physicist who had a powerful intuition. Ernst Mach made major contributions to physics, philosophy, and physiological psychology. In physics, the speed of sound bears his name, as he was the first to systematically study supersonic motion. He also made important contributions to understanding the Doppler Effect. His critique of Newtonian ideas of absolute space and time was an inspiration to the young Einstein, who credited Mach as being the philosophical forerunner of relativity theory. His systematic skepticism of the old physics was similarly important to a generation of young German physicists. He envisioned the inertia of a body (the tendency of a body at rest to remain at rest and of a body in motion to continue moving in the same direction) as the consequence of a relationship of that body with the rest of the matter in the universe. He thought that each body in the universe interacts with all the other bodies in the universe, even at gigantic distances, and its inertia is the sum of those myriad interactions. Figure 2.4 shows a photograph of a bullet in supersonic flight, published by Ernst Mach in 1887.

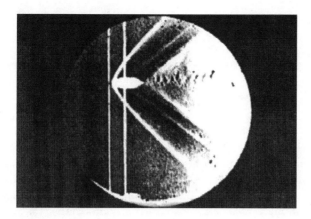

***Figure* 2.4** Photograph of a bullet in supersonic flight, published by Ernst Mach in 1887. (Courtesy of National Aeronautics and Space Administration, http://history .nasa.gov/SP-4219/4219-065.jpg.)

In philosophy, Ernst Mach is best known for his influence on the Vienna Circle (a predecessor of which was named the Ernst Mach Verein), for his famous anti-metaphysical attitude (which developed into the verifiability theory of meaning), for his anti-realist stance in opposition to atomism, and in general for his positivist-empiricist approach to epistemology. It is important to note that some of these influences are currently being re-examined, and are now thought to be both more tenuous and more complicated than was once assumed. He was also an important historian of science, and occupied the Chair for the Philosophy of the Inductive Sciences at the University of Vienna. Although previous philosophers had commented on science and many scientists had influenced philosophy, Mach more than anyone else bridged the divide; he is a founder of the philosophy of science.

In psychology, he studied the relationship of our sensations to external stimuli. Space, time, color, sound, once the domain of physics, were now also being studied by psychologists and conceived of as not only the stuff of the physical external world but also the elements of our inner experience. Mach was deeply inspired by Gustav Fechner's psychophysics here. Psychologists today regard him as a founder of Gestalt theory as well as the discoverer of neural inhibition. Importantly, although in the twentieth century he was better known to philosophers for his influence on physics and the philosophy of physics, it was psychology that was the primary driving force behind his philosophy of science.

Ernst Mach performed pioneering research on vestibular function 100 years ago. His experiments were mainly psychophysical and included measurements of threshold and study of the vestibular-visual interaction.

Contrary to general belief, he concluded that the adequate stimulus for the semicircular canals must be pressure. He presented evidence specifically against the sustained endolymph flow theory of Breuer (1874) and Crum Brown (1874), with which he is frequently associated. Ernst Mach is most closely associated with a positivism that demanded a language of close contact with reality. Mach linked this view with the tradition of the quest for an ideal language in which meaning is a property of a word.

The last of the major ideas in physics before relativity came from Henri Poincare, who pioneered "chaos" theory when he pointed out that a slight change in the initial conditions of some equations results in large-scale differences. Some systems live "at the edge": a slight change in the initial conditions can have catastrophic effects on their behavior. The intellectual leadership, though, was passing to the mathematicians.

2.9   Mathematics and philosophy: Intellectual leadership for inventions

By inventing "set theory," Georg Cantor emancipated mathematics from its traditional domain, numbers. He also introduced "numbers" to deal with infinite quantities, "transfinite numbers," because he realized that space and time are made of infinite points, and that, between any two points, there exists always an infinite number of points. Nonetheless, an infinite series of numbers can have a finite sum. These were, after all, the same notions that, centuries before Cantor, had puzzled Zeno. Cantor gave them mathematical legitimacy.

Gottlob Frege (1884) aimed at removing intuition from arithmetic. He thus set out, just like Leibniz and Boole before him, to replacing natural language with the language of logic, "predicate calculus." Extending Cantor's program, Frege turned mathematics itself into a branch of logic: using Cantor's "sets," he reconstructed the cardinal numbers by a purely logical method that did not rely on intuition.

Frege realized that logic was about the "syntax," not the "semantics" of propositions. An expression has the following two elements:

1. "Sense" or intension, and,
2. "Reference" or extension.

As an instance, "Red" is the word for the concept of redness and the word for all the things that are red. In some cases, expressions with different senses actually have the same referent. For example, "the star of the morning" and "the star of the evening" both refer to Venus. In particular, propositions of logic can have many senses, but only have one of two referents: true or false. Giuseppe Peano was pursuing a similar program at the same time, an "axiomatization" of the theory of natural numbers.

Charles Peirce gave a pragmatic definition of "truth": something is true if it can be used and validated. Thus, truth is defined by consensus. Truth is not agreement with reality, it is agreement among humans. Truth is "true enough." Truth is not eternal. Truth is a process, a process of self-verification. In general, he believed that an object is defined by the effects of its use: a definition that works well is a good definition. An object "is" its behavior. The meaning of a concept consists in its practical effects on our daily lives: if two ideas have the same practical effects on us, they have the same meaning.

Peirce was therefore more interested in "beliefs" than in "truths." Beliefs lead to habits that get reinforced through experience. He saw that the process of habit creation is pervasive in nature: all matter can be said to acquire habits, except that the "beliefs" of inert matter have been fixed to the extent that they can't be changed anymore. Habit is, ultimately, what makes objects what they are. It is also what makes us what we are: I am my habits. Habits are progressively removing chance from the universe. The universe is evolving from an original chaos in which chance prevailed and there were no habits toward an absolute order in which all habits have been fixed.

At the same time, Peirce realized that Frege's theory of sense and referent was limited. Peirce introduced the first version of "semiotics" that focused on what signs are:

- An index is a sign that bears a causal relation with its referent. For example, cigarette smoke "means" that someone was in the same location.
- An icon is a sign that bears a relation of similarity with its referent. For example, the image of a car refers to the car.
- A symbol is a sign that bears a relation with its referent that is purely conventional. For example, the letters "car" refer to a car.

A sign refers to an object only through the mediation of other signs. There is an infinite regression of signs from the signifier, the sign, to the signified, the referent. A dictionary defines a word in terms of other words that are defined in terms of other words, which are defined in terms of other words, and so forth. "Semiosis" means making signs. Peirce believed that knowing is "semiosis." He considered semiosis an endless process.

Philosophy was less interested in logic and more interested in the human condition, the "existentialist" direction that Schopenhauer and Kierkegaard had inaugurated. Friedrich Nietzsche believed that humans are driven by the "will to power," an irresistible urge to order the course of one's experiences. This an extension of Schopenhauer's will to live.

All living beings strive for a higher order of their living condition to overcome their present state's limitations. Human limitations are exemplified by science: science is only an interpretation of the world. Truth and knowledge are only relative to how useful they are to our "will to power."

Henri Bergson was, instead, a very spiritual philosopher, for whom reality was merely the eternal flow of a pantheistic whole. This flow has two directions: the upward flow is life; the downward flow is inert matter. Humans are torn between intellect and intuition:

- Intellect is life observing inert matter in space. Intellect can "understand" inert matter, not only intuition can "grasp" life. In order to understand matter, intellect breaks it down into objects located in space.
- Intuition is life observing life (in time). In contrast to intellect, intuition grasps the flow of life as a whole in time.

2.10   The world is a fiction, a product of the mind

Francis-Herbert Bradley was the last major "idealist." He argued that all categories of science (e.g., space and time) can be proven to be contradictory. This proves that the world is a fiction, a product of the mind. The only reality has to be a unity of all things, the absolute.

Inevitably, the focus of knowledge shifted toward the psyche. William James adapted Peirce's "pragmatism" to the realm of the mind. He believed that the function of the mind is to help the body to live in an environment, just like any other organ. The brain is an organ that evolved because of its usefulness for survival. The brain is organized as an associative network, and associations are governed by a rule of reinforcement, so that it creates "habits" out of regularities, stimulus-response patterns. A habit gets reinforced as it succeeds. The function of thinking is pragmatic: to produce habits of action. James was intrigued by the fact that the brain, in doing so, also produced "consciousness," and considered that mental life is not a substance; instead, it is a process of "the stream of consciousness."

Edward Thorndike postulated the "law of effect": animals learn based on the outcome of their actions. He envisioned the brain as a network: learning occurs when elements are connected. Behavior is due to the association of stimuli with responses that is generated through those connections. A habit is a chain of "stimulus-response" pairs.

Wilhelm-Max Wundt had founded psychology to study the psyche via experiments and logic, not mere speculation. The classical model of psychology was roughly this. Actions have a motive. Motives are hosted in our minds and controlled by our minds. Motives express an imbalance between desire and reality that the mind tries to remedy by changing the reality via action. An action, therefore, is meant to restore the balance between reality and our desires.

What about dreams? Sigmund Freud applied the classical model of psychology. He decided that dreams have a motive, that those motives are in the mind, and that they are meant to remedy an imbalance. Except that the motives of dreams are not conscious: the mind contains both conscious motives and unconscious motives. There is a repertory of motives that our mind, independent of our will, has created over the years, and they participate daily in determining our actions. Freud's revolution was in separating motive and awareness. A dream is only apparently meaningless: it is meaningless if interpreted from the conscious motives. Actually the dream is perfectly logical if one considers also the unconscious motives. The meaning of dreams is hidden and reflects memories of emotionally meaningful experience. Dreams are not prophecies, as ancient oracles believed, but hidden memories. Psychoanalysis was the discipline invented by Freud to sort out the unconscious mess.

Freud divided the self in different parts that coexist. The ego perceives, learns, and acts consciously. Being largely unconscious, the superego is the moral conscience that was created during childhood by parental guidance as an instrument of self-repression. This results in the repertory of unconscious memories created by "libido." He believed that the main motive was "libido." Here, Freud painted a repulsive picture of the human soul.

Carl Jung shifted the focus toward a different kind of unconscious, the collective unconscious. He saw motives not so much in the history of the individual as in the history of the entire human race. His unconscious is a repertory of motives created over the millennia and shared by all humankind. Its "archetypes" spontaneously emerge in all minds. All human brains are "wired" to create some myths rather than others. Thus, mythology is the key to understanding the human mind because myths are precisely the keys to unlock those motives. Dreams reflect this collective unconscious, and therefore connect the individual with the rest of humankind and its archaic past. For Jung, the goal of psychoanalysis is a spiritual renewal through the mystical connection with our primitive ancestors.

Another discipline invented at the turn of the century was hermeneutics. Wilhelm Dilthey argued that human knowledge can only be understood by placing the knower's life in its historical context. Understanding a text implies understanding the relationship between the author and its age. This applies in general to all cultural products, because they are all analogous to written texts.

Ferdinand Saussure was the father of "structuralism." The meaning of any human phenomenon (e.g., language) lays out the network of relationships of which it is part. A sign is meaningful only within the entire network of signs, and the meaning of a sign "is" its relationship to other signs. Language is a system of signs having no reference to anything outside itself. Saussure distinguishes between the following:

- "Parole": a specific utterance in a language, or a speaker's performance, and
- "Langue": the entire body of the language, or a speaker's competence.

Separating "parole" from "langue," Saussure creates the foundations for linguistics.

Edmund Husserl's aim was to found "phenomenology," the science of phenomena. He believed that the essence of events is not their physical description provided by science, but the way we experience them. In fact, science caused a crisis by denying humans the truth of what they experience, by moving away from phenomena as they are. He pointed out that consciousness is "consciousness of":

- Knowing ("noesis"), and
- Known ("noema").

Consciousness correlates the act of knowing ("noesis") of the subject and the object that is known ("noema"). The self knows a phenomenon "intuitively." The essence ("eidos") of a phenomenon is the sum of all possible "intuitive" ways of knowing that phenomenon. The eidos can be achieved only after "bracketing out" the physical description of the phenomenon, only after removing the pollution of science from the human experience, so that the self can experience a purely transcendental knowledge of the phenomenon. This would restore the unity of subject and object that science separated.

2.11   The invention of new physics: Foundation of modern industrial design engineering

In physics, a number of ideas were converging toward the same view of the world. Henri Poincare showed that the speed of light has to be the maximum speed and that mass depends on speed. Hendrik Lorentz unified Newton's equations for the dynamics of bodies and Maxwell's equations for the dynamics of electromagnetic waves in one set of equations, the "Lorentz transformations." These equations, which were hard to dispute because both Newton's and Maxwell's theories were confirmed by countless experiments, contained a couple of odd implications: bodies seemed to contract with speed, while clocks seemed to slow down.

Albert Einstein devised an elegant unification of all these ideas that matched, in scope; the one provided two centuries earlier by Newton. He used strict logic. His axioms were that the laws of nature must be uniform, that those laws must be the same in all frames of reference that are "inertial" (at rest or moving of linear uniform motion), and that the speed of light was the same in all directions. He took the oddities of the Lorentz transformations literally: length and duration appear different to different observers, depending on their state of motion, because space and time are relative. "Now" and "here" became meaningless concepts. The implications of his axioms were even more powerful. All physical quantities were now expressed in four dimensions, a time component and a three-dimensional space component. One, in particular, represented both energy and momentum, depending on the space-time coordinate that one examined. It also yielded the equivalence between mass and energy ($E = mc^2$). Time does not flow (no more than space does): it is just a dimension. A life is a series of points in space-time, points that have both a spatial and a temporal component.

Einstein's world was still Newton's world, though, in some fundamental ways. For example, it was deterministic: the past determines the future. There was one major limitation: because nothing can travel faster than light, there is a limit to what can happen in one's life. Each observer's history is constrained by a cone of light within the space-time continuum radiating from the point (space and time) where the observer "is."

Einstein's next step was to look for a science that was not limited to "inertial" systems. He believed that phenomena should appear the same for all systems accelerated with respect to one another. His new formulas had new startling implications. The dynamic of the universe was reduced to the interaction between masses and the geometry of space-time: masses curve space-time, and the curvature of space-time determines how masses move. Space-time is warped by all the masses with which it is studded. Every object left to itself moves along a "geodesic" of space-time (the shortest route between two points on the warped surface of space-time). It also happens that space-time is warped, and thus objects appear to be "attracted" by the objects in space-time that have warped it. But each object is simply moving on a geodesic (the equivalent of a straight line in traditional "flat" space). It is space-time that is curved, not the geodesic (the trajectory) of the body. Space-time "is" the gravitational field. Einstein thus reduced physics to geometry. The curvature of space-time is measured by a "curvature tensor" (as in Riemann's geometry) such that each point in space-time is described by 10 numbers (the "metric tensor"). If the metric tensor is reduced to zero curvature, one obtains traditional physics in traditional flat space. Curvature (i.e., a gravitational field) also causes clocks to slow down and light to be deflected.

Time and space, according to Einstein's theories of relativity, are woven together, forming a four-dimensional fabric called "space-time." The tremendous mass of Earth dimples this fabric, much like a heavy person sitting in the middle of a trampoline. Gravity, says Einstein, is simply the motion of objects following the curvaceous lines of the dimple. Einstein realized that the traditional gravitational field can be understood as the motion of particles—stars, planets, and even light—on the stretched and curved surface of space-time (see Figure 2.5).

Surprisingly, Einstein's relativity, which granted a special status to the observer, reopened the doors to Eastern spirituality. Nishida Kitaro was perhaps the most distinguished Eastern philosopher to attempt a unification of Western science and Zen Buddhism. In Kitaro's system, Western science is like a robot without feelings or ethics that provides the rational foundations for life, while Zen provides the feelings and the ethics. "Mu" is the immeasurable moment in space-time ("less than a moment") that has to be "lived" in order to reach the next "mu." The flow of "mu" creates a space-time topology. Mu's infinitesimal brief presence creates past, present, and future. The "eternal now" contains one's whole being and also the being of all other things. The present is merely an aspect of the eternal. The eternal generates all the time at present. Mu also creates self-consciousness and free will. There is a fundamental unity of the universe, in particular between the self and the world. Each self and each thing are expressions of the same reality, God. The self is not a substance: it is nothingness ("to study the self is to forget the self"). Religion, not science, is the culmination of knowledge. It is also the culmination of love.

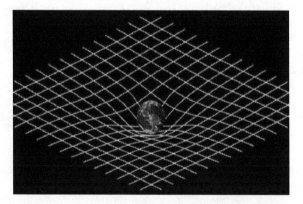

Figure 2.5 Gravity can be thought of as the movements of particles through curved space-time. (Courtesy of National Aeronautics and Space Administration, http://asd.gsfc.nasa.gov/blueshift/wp-content/uploads/2015/11/Spacetime _curvature.png.)

2.12   Unprecedented boom in literature and technology

The European countries (and at least two of their former colonies, Brazil and the United States) experienced an unprecedented boom in literature. The great novels of the time expanded over the genres invented by the previous generations:

- Leo Tolstoy's *War and Peace* (1869)
- George Eliot's *Middlemarch* (1872)
- Emile Zola's *L'Assommoir* (1877)
- Fodor Dostoevsky's *Brothers Karamazov* (1880)
- Joaquim-Maria Machado de Assis' *Memorias Postumas* (1881)
- Joris Huysmans' *A Rebours* (1884)
- Perez Galdos' *Tristana* (1892)
- Jose-Maria Eca de Queiros' *Casa de Ramires* (1897)
- Thomas Mann's *Buddenbrooks* (1901)
- Henry James' *Golden Bowl* (1904)
- Luigi Pirandello's *Il Fu Mattia Pascal* (1904)
- Joseph Conrad's *Nostromo* (1904)
- Maksim Gorkij's *The Mother* (1907)
- Franz Kafka's *Der Prozess* (1915)

Theatre was largely reinvented both as a realist and as a fantastic art through:

- Henrik Ibsen's "Wild Duck" (1884)
- Alfred Jarry's "Ubu Roi" (1894)
- August Strindberg's "The Dream" (1902)
- Anton Chekhov's "The Cherries Garden" (1904)
- Gerhart Hauptmann's "The Weavers" (1892)
- Arthur Schnitzler's "Reigen" (1896)
- Frank Wedekind's "The Book of Pandora" (1904)
- Bernard Shaw's "Pygmalion" (1914)

Poetry works outside of France's "isms" included:

- Robert Browning's "The Ring and the Book" (1869)
- Gerald-Manley Hopkins' "The Wreck of the Deutschland" (1876)
- Ruben Dario's "Prosas Profanas" (1896)
- Giovanni Pascoli's "Canti di Castelvecchio" (1903)
- Antonio Machado's "Campos de Castilla" (1912)
- Rabindranath Tagore's "Gitanjali" (1913)

In the new century, France still led the way of literary fashion with Guillaume Apollinaire's "Alcools" (1913) and Paul Valery's "La Jeune Parque" (1917).

Classical music reflected the nationalist spirit of the age, including:

- Richard Wagner in Germany
- Hector Berlioz in France, Modest Moussorgsky in Russia
- Giuseppe Verdi in Italy
- Antonin Dvorak in the Czech Republic
- Fryderyk Chopin in Poland
- Ferencz Liszt in Hungary
- Johannes Brahms, Richard Strauss, Joseph Bruckner and Gustav Mahler in the German-speaking world (Austria and Germany, etc.)

At the beginning of the new century, a number of compositions announced that the classical format was about to exhaust its mission:

- Aleksandr Skrjabin's "Divine Poem" (1905)
- Arnold Schoenberg's "Pierrot Lunaire" (1912)
- Claude Debussy's "Jeux" (1912)
- Igor Stravinskij's "Le Sacre du Printemps" (1913)
- Charles Ives' "Symphony 4" (1916)
- Sergej Prokofev's "Classic Symphony" (1917)
- Erik Satie's "Socrates" (1918)

Niels Bohr (1913) showed that electrons are arranged in concentric shells outside the nucleus of the atom, with the number of electrons determining the atomic number of the atom and the outermost shell of electrons determining its chemical behavior. Paul Rutherford (1919) showed that the nucleus of the atom contains positively charged particles (protons) in equal number to the number of electrons. In 1932, James Chadwick showed that the nucleus of the atom contains electrically neutral particles (neutrons): isotopes are atoms of the same element (containing the same number of electrons/protons) but with different numbers of neutrons. Their model of the atom was another case of nature preferring only discrete values instead of all possible values.

At this point, physics was aware of three fundamental forces:

- The electromagnetic force
- The gravitational force
- The nuclear force

The theory that developed from these discoveries was labeled "quantum mechanics." It was born to explain why nature prefers some "quanta" instead of all possible values. Forces are due to exchanges of discrete amounts of energy ("quanta").

The key intuition came in 1923, when Louis DeBroglie argued that matter can be viewed both as particles and waves: they are dual aspects of the same reality. This also explained the energy-frequency equivalence discovered by Albert Einstein in 1905: the energy of a photon is proportional to the frequency of the radiation.

Max Born realized (1926) that the "wave" corresponding to a particle was a wave of probabilities; it was a representation of the state of the particle. Unlike a pointless particle, a wave can be in several places at the same time. The implication was that the state of a particle was not a specific value, but a range of values. A "wave function" specifies the values that a certain quantity can assume, and, in a sense, states that the quantity *has* all those values (e.g., the particle *is* in all the places compatible with its wave function). The "wave function" summarizes ("superposes") all the possible alternatives. Erwin Schroedinger's equation describes how this wave function evolves in time, just like Newton's equations describe how a classical physical quantity evolves in time. The difference is that, at every point in time, Schroedinger's equation yields a range of values (the wave function), not a specific value.

The probability associated with each of those possible values is the probability that an observation would reveal that specific value (e.g., that an observation would find the particle in one specific point). This was a dramatic departure for physics. Determinism was gone because the state of a quantum system cannot be determined anymore. Chance had entered the picture because, when a physicist performs an observation, nature decides randomly which of the possible values to reveal. And a discontinuity had been introduced between unobserved reality and observed reality: as long as nobody measures it, a quantity has many values (e.g., a particle is in many places at the same time), but, as soon as someone measures it, the quantity assumes only one of those values (e.g., the particle is in one specific point).

The fact that the equations of different quantities were linked together (a consequence of Einstein's energy-frequency equivalence) had another odd implication, expressed by Werner Heisenberg's "uncertainty principle": there is a limit to the precision with which we can measure quantities. The more precise we want to be about a certain quantity, the less precise we will be about some other quantity.

Space-time turns out to be discrete: there is a minimum size to lengths and intervals, below which physics ceases to operate. Thus, there is a limit to how small a physical system can be. Later, physicists would realize that a vacuum itself is unrecognizable in quantum mechanics: it is not empty.

Besides randomness, which was already difficult to digest, physicists also had to accept "non-locality": a system can affect a distant system despite the fact that they are not communicating. If two systems get entangled in a wave, they will remain so forever, even if they move to the opposite sides of the universe, at a distance at which a signal cannot travel in time to tell one what the other one is doing.

If this were not enough, Paul Dirac (1928) realized that the equations of quantum mechanics allowed for "anti-matter" to exist next to usual matter, for example, a positively charged electron exists that looks just like the electron but has the opposite charge. Paul Dirac's equations for the electron in an electromagnetic field, which combined quantum mechanics and special relativity, transferred quantum theory outside mechanics, into quantum electrodynamics.

Perhaps the most intriguing aspect of quantum mechanics is that a measurement causes a "collapse" of the wave function. The observer changes the course of the universe by the simple act of looking at a particle inside a microscope.

This led to different "interpretations" of quantum mechanics. Niels Bohr argued that maybe only phenomena are real. Werner Heisenberg, instead, thought that maybe the world is made of possibility waves. Paul Dirac thought that quantum mechanics simply represents our (imperfect) knowledge of a system. Hugh Everett took the multiple possible values of each quantity literally, and hypothesized that we live in an ever multiplying "multiverse": at each point in time, the universe splits according to all the possible values of a measurement. In each new universe one of the possible values is observed, and life goes on.

John Von Neumann asked at which point the collapse occurs. If a measurement causes nature to choose one value, and only one, among the many that are allowed by Schroedinger's equation, "when" does this occur? In other words, where in the measuring apparatus does this occur? The measurement is performed by having a machine interact with the quantum system and eventually deliver a visual measurement to the human brain. Somewhere in this process a range of possibilities collapses into one specific value. Somewhere in this process the quantum world of waves collapses into the classical world of objects. Measurement consists in a chain of interactions between the apparatus and the system, whereby the states of the apparatus become dependent on the states of the system. Eventually, states of the observer's consciousness are made dependent on states of the system, and the observer "knows" what the value of the observable is. If we proceed backward, this seems to imply that the "collapse" occurs in the conscious being, and that consciousness creates reality.

Einstein was the main critic: he believed that quantum mechanics was an incomplete description of the universe, and that some "hidden variables" would eventually turn it into a deterministic science just like traditional science and his own relativity.

From the beginning, it was obvious what was going to be the biggest challenge for quantum mechanics: discovering the "quantum" of gravitation. Einstein had explained gravitation as the curvature of space-time, but quantum mechanics was founded on the premise that each force is due to the exchange of quanta: Gravity did not seem to work that way, though.

A further blow to the traditional view of the universe came from Edwin Hubble's discovery (1929) that the universe is expanding. It is not only the Earth that is moving around the Sun, and the Sun that is moving around the center of our galaxy, but also all galaxies are moving away from each other.

The emerging discipline was biology. By the 1940s, Darwin's theory of evolution (variation plus selection) had been finally wed to Mendel's theory of genetic transmission (mutation), yielding the "modern synthesis." Basically, Mendel's mutation explained where Darwin's variation came from. At the same time, biologists focused on population, not individuals, using the mathematical tool of probabilities. "Population Genetics" was born.

Erwin Schroedinger noticed an apparent paradox in the biological world: as species evolve and as organisms grow, life creates order from disorder, thus contradicting the second law of thermodynamics. The solution to this paradox is that life is not a "closed" system: the biological world is a world of energy flux. An organism stays alive (i.e., maintains its highly organized state) by absorbing energy from the outside world and processing it to decrease its own entropy (i.e., increase its own order). "Living organisms feed upon negative entropy." Life is "negentropic." The effect of life's negentropy is that entropy increases in the outside world. The survival of a living being depends on increasing the entropy of the rest of the universe.

Erwin Schroedinger also noticed an apparent paradox in the physical world: A cat is placed in a box, together with a radioactive atom. If the atom decays, and the geiger-counter detects an alpha particle, the hammer hits a flask of prussic acid (HCN), killing the cat. The paradox lies in the clever coupling of quantum and classical domains. Before the observer opens the box, the cat's fate is tied to the wave function of the atom, which is itself in a superposition of decayed and undecayed states. Thus, said Schroedinger, the cat must itself be in a superposition of dead and alive states before the observer opens the box, "observes" the cat, and "collapses" its wave function (see Figure 2.6).

However, the lives of ordinary people were probably more affected by a humbler kind of science that became pervasive: synthetic materials. In 1907, Leo Baekeland invented the first plastic ("bakelite"). In 1925, cellophane was introduced and in 1930 it was the turn of polystyrene. In 1935, Wallace Carothers invented nylon.

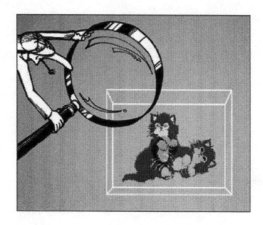

Figure 2.6 Schroedinger's cat. (Courtesy of Science@Berkeley Lab, http://www2 .lbl.gov/Science-Articles/Archive/sabl/2005/June/assets/img/02_Schrodinger's _cat.jpg.)

The influence of Einstein can also be seen on Samuel Alexander, who believed in "emergent evolution": existence is hierarchically arranged and each stage emerges from the previous one. Matter emerges from space-time, life emerges from matter, mind emerges from life, and so on.

Arguing against idealism, materialism, and dualism, Bertrand Russell took Einstein literally and adopted the view that there is no substance ("neutral monism"): everything in the universe is made of space-time events, and events are neither mental nor physical. Matter and mind are different ways of organizing space-time. Elsewhere, he conceived of consciousness as a sense organ that allows us to perceive some of the processes that occur in our brain. Consciousness provides us with direct, immediate awareness of what is in the brain, whereas the senses "observe" what is in the brain. What a neurophysiologist really sees while examining someone else's brain is part of his or her own (the neurologist's) brain.

2.13   *Paradox and influence on scientific thinking*

Russell's paradox is the most famous of the logical or set-theoretical paradoxes. Also known as the Russell-Zermelo paradox, the paradox arises within naïve set theory by considering the set of all sets that are not members of themselves. Such a set appears to be a member of itself if and only if it is not a member of itself. Hence, the paradox.

Some sets, such as the set of all teacups, are not members of themselves. Other sets, such as the set of all non-teacups, are members of themselves. Call the set of all sets that are not members of themselves "R." If R is a member of itself, then by definition it must not be a member of itself.

Similarly, if R is not a member of itself, then by definition it must be a member of itself.

Although also noticed by Ernst Zermelo, the contradiction was not thought to be important until it was discovered independently by Bertrand Russell in the spring of 1901. Since then, the paradox has prompted a great deal of work in logic, set theory, and the philosophy and foundations of mathematics.

Bertrand Russell was perhaps more influential in criticizing Frege's program. Over the course of his life, Gottlob Frege formulated two logical systems in his attempts to define certain basic concepts of mathematics and to derive certain mathematical laws from the laws of logic. In his book of 1879, *Begriffsschrift: eine der arithmetischen nachgebildete Formelsprache des reinen Denkens* (an arithmetic simulated formula of the pure thought), he developed a second-order predicate calculus and used it both to define interesting mathematical concepts and to state and prove mathematically interesting propositions. However, in his two-volume work of 1893/1903, *Grundgesetze der Arithmetik* (Basic Laws of Arithmetic), Frege added (as an axiom) what he thought was a logical proposition (Basic Law V) and tried to derive the fundamental axioms and theorems of number theory from the resulting system.

However, Bertrand Russell found a paradox that seemed to terminate the program to formalize mathematics: the class of all the classes that are not members of themselves is both a member and not a member of itself. Russell's Paradox can be put into everyday language in many ways. The most often repeated is the Barber Question. It goes like this:

> In a small town there is only one barber. This man is defined to be the one who shaves all the men who do not shave themselves. The question is then asked,
>   'Who shaves the barber?'
>   If the barber doesn't shave himself, then—by definition—he does.
>   And, if the barber does shave himself, then—by definition—he does not.

As shown in Figure 2.7, another popular form of Russell's Paradox is the following:

> This statement is false.
>   If the statement is false, then it is true; and if the statement is true, then it is false.

Bertrand Russell solved the paradox (and other similar paradoxes, such as the proposition "I am lying" which is true if it is false and false if it is true) by introducing a "theory of types," which basically resolved

```
Consider the statement
'This statement is false.'

If the statement is false,
   then it is true;
and
if the statement is true,
   then it is false.
```

Figure 2.7 Bertrand Russell's Paradox.

logical contradictions at a higher level. What is the primary feature of Bertrand Russell's Paradox? It is one of self-reference. Thus, it is like the Liar Paradox in which a sentence (or speaker) speaks about itself. With Russell's Paradox, it is a case of a set (or a class) referring to itself or including itself as a member of itself.

Ludwig Wittgenstein erected another ambitious logical system. Believing that most philosophical problems are non-issues created by linguistic misunderstandings, he set out to investigate the nature of language. He concluded that the meaning of the world cannot be understood from inside the world, and thus metaphysics cannot be justified from inside the world (no more and no less than religion or magic). Mathematics also lost some of its appeal: it cannot be grounded in the world, therefore it is but a game played by mathematicians.

Wittgenstein saw that language has a function, that words are tools. Language is a game between people, and it involves more than a mere transcription of meaning: it involves assertions, commands, questions, and so on. The meaning of a proposition can only be understood in its context, and the meaning of a word is due to the consensus of a society. To understand a word is to understand a language.

Edward Sapir argued that language and thought influence each other. Thought shapes language, but language also shapes thought. In fact, the structure of a language exerts an influence on the way its speakers understand the world. Each language contains a "hidden metaphysics," an implicit classification of experience, a cultural model, a system of values. Language implies the categories by which its speakers not only communicate but also think.

Lev Vygotsky reached a similar conclusion from a developmental viewpoint: language mediates between society and the child. Language guides the child's cognitive growth. Thus, cognitive faculties are merely internalized versions of social processes that we learned via language as children. Thus, one's cognitive development (way of thinking) depends on the society in which he or she grew up.

Something similar to the wave/particle dualism of physics was taking place in psychology. Behaviorists such as John Watson, Ivan Pavlov, and Burrhus Skinner believed that behavior is due to stimulus-response patterns. Animals learn how to respond to a stimulus based on reward/punishment, that is, via selective reinforcement of random responses. All of behavior can be reduced to such "conditioned" learning. This also provided an elegant parallel with Darwinian evolution, which is also based on selection by the environment of random mutations. Behaviorists downplayed mind: thoughts have no effect on our actions.

Cognitivists such as Max Wertheimer, Wolfgang Kohler, and Karl Lashley (the "gestalt" school) believed just the opposite: an individual stimulus does not cause an individual response. We perceive (and react to) "form," as a whole, not individual stimuli. We recognize objects not by focusing on the details of each image, but by focusing on the image as a whole. We solve problems not by breaking them down in more and more minute details, but via sudden insight, often by restructuring the field of perception. Cognitivists believed that the processing (thought) between input and output was the key to human behavior, whereas Behaviorists believed that behavior was just a matter of linking outputs with inputs.

Cognitivists conceived the brain as a holistic system. Functions are not localized but distributed around the brain. If a piece of the brain stops working, the brain as a whole may still be working. They envisioned memory as an electromagnetic field, and a specific memory as a wave within that field.

Otto Selz was influenced by this school when he argued that to solve a problem entails recognizing the situation and filling in the gaps: information in excess contains the solution. Thus, solving a problem consists in comprehending it, and comprehending it consists in reducing the current situation to a past situation. Once we "comprehend" it, we can also anticipate what comes next: inferring is anticipating.

Last, but not least, Fredrick Bartlett suggested that memory is not a kind of storage, because it obviously does not remember the single words and images. Memory "reconstructs" the past. We are perfectly capable of describing a scene or a novel or a film even though we cannot remember the vast majority of the details. Memory has "encoded" the past in an efficient format of "schemata" that bear little resemblance to the original scenes and stories, but that take little space and make it easy to reconstruct them when needed.

Kurt Goldstein's theory of disease is also an example of cognitivist thinking. Goldstein took issue against dividing an organism into separate "organs": it is the whole that reacts to the environment. A "disease" is the manifestation of a change in the relationship between the organism and its environment. Healing is not a "repair," but an adaptation of the whole organism to the new state. A sick body is, in fact, a system that is undergoing global reorganization.

Jean Piaget focused entirely on the mind, and precisely on the "growth" of the mind. He realized that, during our lifetime, the mind grows, just like the body grows. For him, cognition was self-regulation: organisms need to constantly maintain a state of equilibrium with their environment.

Piaget believed that humans achieve that equilibrium through a number of stages, each stage corresponding with a reorganization of our cognitive life. This was not a linear, gradual process of learning, but a discontinuous process of sudden cognitive jumps. Overall, the growth of the mind was a transition from the stage of early childhood, in which the dominant factor is perception, which is irreversible, to the stage of adulthood in which the dominant factor is abstract thought, which is reversible.

Charlie-Dunbar Broad was a materialist in the age of behaviorists and cognitivists. He believed that the mind was an emergent property of the brain, just like electricity is an emergent property of conductors. Ultimately, all is matter.

That is not to say that the "spiritual" discourse was dead. Martin Buber argued that our original state was one of "I-You," in which the "I" recognizes other "I's" in the world, but we moved toward an "I-It" state, in which the "I" sees both objects and people merely as means to an end. This changes the way in which we engage in dialogue with each other, and thus our existence.

For Martin Heidegger, the fundamental question was the question of "being." A conceptual mistake is to think of the human being as a "what" instead of a "who." Another conceptual mistake is to separate the "who" from the "what": the human being is part of the world, at the same time that it is the observer of the world. The human being is not "Dasein" (existence) but "Dase-in" ("existing in" the world). We cannot detach ourselves from reality because we are part of it. We just don't "act": we are "thrown" in an action. We know what to do because the world is not a world of particles or formulas: it is a world of meaning that the mind can understand. Technology alienates humans because it recasts the natural environment as merely a reservoir of natural resources to be exploited, when in fact it provided them with an identity.

Vladimir Vernadsky introduced the concept of the "biosphere" to express the unity of all life.

Alfred Whitehead believed in the fundamental unity of the world, due to the continuous interaction of its constituents, and that matter and mind were simply different aspects of the one reality, due to the fact that mind is part of the bodily interaction with the world. He thought that every particle is an event having both an "objective" aspect of matter and a "subjective" aspect of experience. Some material compounds, such as the brain, create the illusion that we call "self." But the mental is not exclusive to humans, it is ubiquitous in nature.

The relationship of the self with the external reality was also analyzed by George Herbert Mead, who saw consciousness as, ultimately, a feature in the world, located outside the organism and due to the interaction of the organism with the environment. Consciousness "is" the qualities of the objects that we perceive. Those qualities are perceived the way they are because of the acts that we performed. The world is the result of our actions. It is our acting in the environment that determines what we perceive as objects. Different organisms may perceive different objects. We are actors as well as observers (of the consequences of our actions). Consciousness is not the brain process: the brain process is only the switch that turns consciousness on or off. Consciousness is pervasive in nature. What is unique to humans, as a social species, is that they can report on their conscious experiences. That "reporting" is what we call the "self." A self always belongs to a society of selves.

Sarvepalli Radhakrishnan believed that science was proving a universal process of evolution at different levels (material, organic, biological, social) whose ultimate goal was to reveal the absolute (the spiritual level). Human consciousness is not the last step in evolution, but will be succeeded by the emergence of a super-consciousness capable of realizing the union with a super-human reality that human science cannot grasp.

Muhammad Iqbal believed that humans are imperfect egos who are striving to reach God, the absolute ego.

However, it was an economist, John Maynard Keynes, who framed the fundamental philosophical problem of the postindustrial state. As citizens no longer need to worry about survival, "man will be faced with his real, permanent problem: how to use his freedom."

But Karl Jaspers saw existence as a contradiction in terms. In theory, humans are free to choose the existence they prefer, but in practice it is impossible to transcend the historical and social background. Thus, one is only truly free of accepting of one's destiny. Ultimately, we can only glimpse the essence of our own existence, but we cannot change it.

Several philosophers, particularly the "logical positivists" (such as Rudolf Carnap and Alfred-Jules Ayer), shed new light on this problem. Carnap believed that meaning could be found only in the marriage of science and Frege's symbolic logic. He believed in the motto "the meaning of a proposition is its method of verification," which put all the responsibility on the senses. He demoted philosophy to a second-rate discipline whose only function would be to clarify the "syntax" of the logical-scientific discourse. The problem is that the senses provide a subjective view of the world, and therefore the "meaning" derived from verification is personal, not universal. Also, it is not clear how one can "verify" statements about history. Soon it became clear that even scientific propositions could not quite be "verified" in an absolute way. Last, but not least, Carnap could not prove the very principle of verification based on the principle of verification.

Karl Popper clarified that truth is always and only relative to a theory: no definition of absolute truth is possible. The issue is not what is "true," but what is "scientific." Popper argued that matching the facts was not enough to qualify as "scientific": a scientific theory should also provide the means to falsify itself.

chapter three

Safety, reliability, and risk management

"All the astronauts landing on Mars are engineers; we are bringing them home safely ..."

Engineering is the art of directing the great sources of power in nature for the use and convenience of man.

Thomas Tredgold (1828)

3.1 Optical quantum technologies, enlightening "Star Wars: The Force Awakens"

Star Wars: The Force Awakens, the seventh installment in the main Star Wars film series, was released in 2D, 3D, and IMAX 3D on December 18, 2015, more than 10 years following the release of the franchise's last installment, *Star Wars Episode III: Revenge of the Sith* (2005).

In the stage of *Star Wars*, an American epic space opera depicting the adventures of various characters "a long time ago in a galaxy far, far away," a quantum computer was a computer capable of generating a random number that was 47 digits long for use as a password. For further protection from slicing, one digit in the number would shift one value lower or higher every six standard hours. Slicing is equal to the real life term hacking.

In *Star Wars Death Star*, which covered the later stages of the battle station's construction and follows the actions of a variety of individuals connected to it until its destruction, the quantum computer was used to create a number that protected a personal folder of Atour Riten, a human male commander and chief librarian who served in the Imperial Navy and the Janissariad during the Balduran Civil War and the Galactic Civil War in 0 BBY. The personal folder contained the Death Star plans, the blueprints of the first Death Star, an Imperial superweapon that was the brainchild of Grand Moff Wilhuff Tarkin. Death Star plans were protected by a series of pyrowalls, a form of file protection, and military wards.

Currently, system architectures are explored for quantum computing, quantum metrology, and quantum memory with research work on a variety of materials systems including semiconductor quantum dots, nanomolecules, and defects in crystals. For optical control of electronic states in these materials, energy harvesting in photovoltaic cells relies on the creation and transport of electronic excitations in nanomolecules of semiconductors. Here, quantum mechanics is used to understand how this process can be optimized for efficient conversion of sunlight into electrical energy for powering quantum computing systems. Commercial devices capable of encrypting information in unbreakable codes exist today, thanks to recent quantum optics advances, especially the generation of photon pairs—tiny entangled particles of light.

The research work relies on understanding quantum systems that interact with an environment to develop open quantum computing systems. As shown in Figure 3.1, one of the properties of light exploited within quantum optics is "photon polarization," which is essentially the direction in which the electric field associated with the photon oscillates. A new approach has been based on a micro-ring resonator—a tiny optical cavity with a diameter on the order of tens to hundreds of micrometers—that operates in such a way that energy conservation constraints suppress classical effects while amplifying quantum processes. While a similar suppression of classical effects has been observed in gas vapors and complex micro-structured fibers, this is the first time it has been applied to develop lasers of ultra-high security, integrated quantum photonics, and on-chip quantum optical computing.

For quantum computing, theoretical techniques are being applied to understand the quantum properties of nanomaterials, focusing on quantum computing, quantum communication, and quantum energy harvesting. Here, the fabrication process of the chip is compatible with that

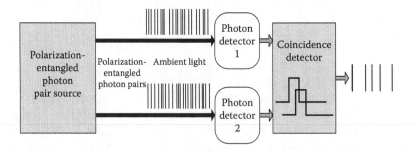

Figure 3.1 A multi-wavelength time-coincident optical communication system. (Courtesy of NASA's Glenn Research Center.)

currently used for electronic chips, enabling future coexistence of quantum devices with standard integrated circuits, which is a fundamental requirement for the widespread adoption of optical quantum technologies, enlightening *Star Wars: The Force Awakens.*

3.2   Applying fault-tolerant quantum computing to mitigate risk and uncertainty

For decades, one of the expeditions of quantum physics has been to build a quantum computer that can process large-scale, challenging computational problems exponentially faster than classical computers. While scientists and engineers are progressing toward this target, almost every part of a quantum computer still needs noteworthy research and development (R&D). Current research is focusing on every angle of the quantum computer problem, including:

- Innovative ways to generate entangled photon pairs,
- Inventive types of gates and their fabrication on chips,
- Superior ways to create and control qubits,
- Novel designs for storage/memory buffers,
- Effective detectors, and
- Creative ways to optimize them in various architectures.

Optimizing the waveguide geometry, integrated quantum optical circuits are constructed to realize single-photon quantum computing. The central elements for such circuits include sources, gates, and detectors. However, a major missing function critical for photonic quantum computing on-chip is a buffer, where single photons are stored for a short period of time to facilitate circuit synchronization. As a significant step in the field, an all-optical integrated quantum processor is being developed at the National Institute of Quantum Computing (NIQC). For fault-tolerant quantum computing, the research explores the frontier of current quests for quantum processing of ultra-high security, integrating the following enabling techniques including:

- Probabilistic Bayesian network,
- Quantum filtering,
- Error correction code (ECC), and
- Riemannian geometry

As shown in Figure 3.2, here is a love story at the smallest scale imaginable: particles of light. It is possible to have particles that are so intimately linked that a change to one affects the other, even when they are separated at a distance.

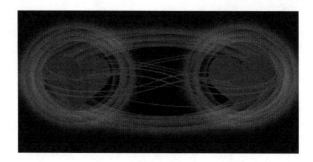

Figure 3.2 Technology used to study the "love" between particles is also being used in research to improve communications between space and Earth. (Credit: NASA/JPL-Caltech.)

This idea, called "entanglement," is part of the branch of physics called quantum mechanics, a description of the way the world works at the level of atoms and particles that are even smaller. Quantum mechanics says that at these very tiny scales, some properties of particles are based entirely on probability. In other words, nothing is certain until it happens.

3.3   Entanglement: How to flip a quantum coin at nanometer scale

"Flip a Quantum Coin at Nanometer Scale: Heads-Up and Tails-Up at the Same Time" described "entanglement: a physical phenomenon that occurs when pairs or groups of particles are generated or interact in ways such that the quantum state of each particle cannot be described independently—instead, a quantum state may be given for the system as a whole. This is like having an entangled pair of our superposition heads/tails block and placing them on opposite sides of the coin. As soon as one of them is observed—let's say it becomes heads—we observe its entangled counterpart and find that it, too, becomes heads." In 1972, Freeman and Clauser succeeded for the first time in preparing two particles that exhibited a strange condition, predicted by quantum theory, called "entanglement." The condition had been discussed theoretically by Einstein and co-workers in 1935, and at that time they argued that, because such a thing was obviously impossible, there must be something wrong with quantum theory. Freeman and Clauser's work, and subsequent more detailed experiments that fully confirmed the prediction of quantum theory, triggered a tide of speculation. There seemed no limit to the mysteries that might now be explained using this new phenomenon: telepathy, consciousness, healing, and so on were all examined. Now that the production of entangled particles has become almost routine technology, it is perhaps a good time to take stock of what we have learned.

But first, what is entanglement? All the observations have been made on very simple microscopic particles. Briefly, the essence of the idea is the productions of pairs of particles which, though separated by a large distance, show correlations in their behavior that are inexplicable on a basis of old (non-quantum) physics. Figure 3.3 helps explain the idea of "entangled particles." Alice and Bob represent photon detectors, which NASA's Jet Propulsion Laboratory and the National Institute of Standards and Technology developed.

Flipping a quantum coin at nanometer scale helps us to visualize many different models that we can use to describe how nanoparticles interact with each other in the quantum world. In fact, flipping a quantum coin represents a two-state system (TSS) in the quantum world, where the two sides of the coin would have two possible quantum states. A quantum state is a state of a quantized system that is described by a set of quantum numbers. A quantum number is a number that expresses the value of some property of a particle that occurs in the quanta.

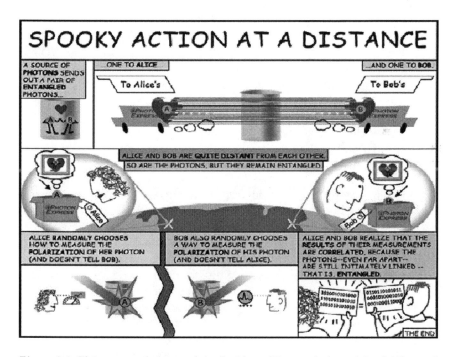

Figure 3.3 This cartoon helps explain the idea of "entangled particles." Alice and Bob represent photon detectors, which NASA's Jet Propulsion Laboratory and the National Institute of Standards and Technology developed. (Credit: NASA/ JPL-Caltech.)

In the quantum world, how do you flip a quantum coin at nanometer scale? Here are different ways:

- Spin a quantum coin. Spin is one of the four basic quantum numbers. It is the intrinsic angular momentum. It defines the spin given to a particle. For a two-state system, spin can exist as counterclockwise and clockwise. It can have a value of either +1/2 or –1/2. This means that no two particles in the same energy level have the same properties or states. Think about the coin, it has a head on one side and an eagle on the other side. There are no two same images per coin. This is the same with spin as a two-level system. One particle has a –1/2 spin while the other particle has a +1/2 spin. Protons, neutrons, electrons, neutrinos, and quarks could all be quantum coins.

- Transition a quantum coin from an excited state to a ground state. Involving photons, this is a quantum system with "atom-light" interaction. Using a quantum coin, you have the excited state on one side and the ground state on the other side. The excited state is where the atom jumps to when energy is added. The ground state is the lowest energy level of the atom. There are two processes that happen between the ground state and the excited state. These processes are absorption and emission. Absorption happens when the atom absorbs a photon, causing the atom to become excited. Emission happens when the atom falls to ground state and releases a photon. With our coin, we can imagine that the coin has been forced to spin or is infinitely flipping. This action demonstrates how absorption and emission are constantly occurring.

- Toss a quantum coin of the ammonia molecule. The nitrogen of ammonia has two molecular states. These states are "up" and "down." Once again, on one side of the coin, you have "up" and on the other, you have "down." These two states are non-degenerate. When something is non-degenerate, it does not have the same quantum energy level. In this situation, when excitation of the molecule happens, vibration is caused by the absorption and re-emission of photons. This is similar to tossing a slinky back and forth in your hand. This quantum phenomena allows the ammonia molecule to have its pyramidal shape and allows ammonia to be used as a source for a special type of laser called microwave amplification of stimulated emission of radiation (MASTER).

- Qubit. Quantum coin for quantum computing. The qubit is used in quantum computing. Like the bit that is used in regular computing, the qubit is the unit of quantum information used in quantum computing. Unlike the bit, the qubit can have a 0 and 1 at the same time. A common example of the two states used in the qubit is polarization. On one side of the coin, there is vertical polarization and on the other, horizontal polarization. You have the value of 0 and perhaps horizontal polarization. On the other side, you have the value of 1 and vertical polarization. The qubit reveals an interesting property about our quantum coin. This property is called superstition entanglement, meaning that two states exist at the same time. Entanglement is when collective properties are shared. In this case, the collective or common property is polarization: vertical and horizontal.

- The doublet. Quantum coin of rotational symmetry. Doublets are spectral lines of an ionized gas that have been split into two lines under the influence of a magnetic field. The doublet would have +1/2 on one side of the coin and –1/2 on the other side of the coin. The doublet reveals another unique feature about our quantum coin. This feature is called rotational symmetry. This means that, regardless of how you rotate the coin, the value is still 1/2.

Flipping a quantum coin, the concept of the two-state quantum system is being applied to optimize the flow of quantum information, mitigating risk and uncertainty of nano-manufacturing.

Entanglement experiments use a property of light called polarization, to do with the direction in which the fields that constitute light are vibrating. (Polaroid sunglasses filter out light with a particular direction of polarization.) It is possible, using a special optical material, to split a single photon into two so-called daughter photons. These two are allowed to travel apart (by more than 10 km in some experiments) and then the directions of polarization of the two photons are measured simultaneously. Two points emerge from analyzing the results:

1. The direction of the polarization of either particle is not fully determined before the measurement takes place; it must involve a partly random response of the particle to the measuring apparatus.
2. There is a correlation between the results of the measurements on the two particles. For example (and depending on the arrangement of the measuring apparatus), it might be that if particle A is measured to have its polarization pointing vertically, it is then more likely that the same result will be obtained for particle B.

3.4   Flip a quantum coin at nanometer scale: Heads-up and tails-up at the same time

The first of the two premises key to this conclusion is that space-time is a real entity, an object in the same way particles are objects. An objective space-time would not have a "location" in space-time (there need not be any "other" space-time in which space-time resides), but otherwise it would be a thing, subject itself to the laws of relativity and quantum mechanics. This is not a popular theory, but it has yet to be contradicted by any experiment or observation, and it is fully consistent with general relativity. And there are two observations that support it. However, could it be that the particles are, as it were, preprogrammed when they are split to respond in this way? For example, it might be that the split always results either in both particles vibrating horizontally, and both vibrating vertically. A detailed argument by the renowned theorist of foundational physics John Bell demonstrated that no "preprogramming" could explain the observed results. In other words, the particles were responding spontaneously, but in an interconnected manner. In the everyday, large-scale world, we tend to think things are one way or the other. For example, if I flip a coin, it is either heads or tails. This deterministic nature provides the basis of today's electronics design and green electronics manufacturing.

Quantum theory presents a very precise, and by now almost universally accepted, mathematical account of what is happening, in which entanglement corresponds to a particular mathematical form for the expression describing the pair of particles. What are the possible translations of this mathematics into words and pictures? Quantum engineering indicates that both energy and information are not continuous but instead come in small "packets" known as quantum and that the reality is really probabilistic instead of deterministic.

On large scales, the probabilistic consequences of quantum engineering are averaged out and, therefore, undetectable. However, at the nanometer (10^{-9} m) scale, quantum engineering cannot be ignored and in fact begins to dominate for electronic design. Figure 3.4 shows an experimental

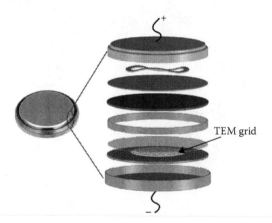

Figure 3.4 An experimental coin cell setup at Brookhaven National Laboratories. (Courtesy of Brookhaven National Laboratories, https://www.bnl.gov/today/body _pics/2014/05/coin-cell-hr.jpg.)

coin cell setup at Brookhaven National Laboratories (BNL). A carbon supported transmission electron microscopy (TEM) grid loaded with a small amount of the nickel-oxide material was pressed against the bulk anode and submerged in the same electrolyte environment.

Let's describe some of the key assertions of the conventional verbal translation of the quantum mechanical account.

1. The properties of particles, properties which are the objects of experimental investigation, do not exist independently of the observation. Rather, they arise in the process of the interaction between the particle and the experimental apparatus. It is even misleading to think of them as "properties of particles" at all: they are aspects of an event of measurement.
2. Spatial separation, and to some extent separation in time, are irrelevant to the correlations produced by entanglement. Spatial and temporal relations do not enter into the calculations at all; the particles could be anywhere.
3. Entanglement is the general rule; any interaction at any time in the past will entangle particles, so that very special conditions have to hold in order to produce particles that are not entangled. The achievement of Clauser and others actually lay not in the mere fact of producing entanglement, but in producing an entanglement that was of such a form that it could be examined experimentally. This point will be crucial below when I come to discuss the wider implications of this work.

Points 1 and 2 here carry an important philosophical message that challenges how we normally think about the universe, which is in terms of definite and separated things located in space. Point 1 undermines the definiteness of "things." It is not saying that physical entities are merely figments of our cultural assumptions (though this may indeed be the case): physicists behave as if they are dealing with what might in some sense be called "reality." But this "reality" is what the philosopher of physics D'Espagnat called veiled. What we experience, either in the artificial setting of a laboratory or in normal moment-to-moment life, is quite distinct from what physicists regard as the foundation of the material universe, namely the abstract entities called particles and fields. And I should add that, while the connection between particles and experience is clear in the case of the laboratory, it remains in many respects obscure and controversial at the level of ordinary life.

It is point 3 that is vital for the wider implications of this subject. On the face of it, it would seem, for example, that we could use pairs of entangled particles for an instant communication system that operated independently of distance—something that would be highly reminiscent of telepathy. (Some authors have even written of one particle "instanta-neously changing its state" when the other is measured, for which there is no justification at all.) More generally, it suggests that the world, rather than being a collection of isolated particles pushing each other around, is more like an intricate web of subtle interconnections. But how far can we take this picture?

The fusion of nanotechnology and quantum engineering leads to the design of new types of nanodevices and nanoscale materials, where func-tionality and structure of quantum nanodevices are described through quantum phenomena and principles such as observational dependency, superposition, and entanglement, as summarized here:

- Observational dependency: observing certain things on the sub-atomic level (like photons) actually caused them to have different properties. Just like if I flip a quantum coin, when you look at a coin from the top, it is heads and when you look at it from the bottom it is tails.
- Superposition: subatomic particles exist in all of their possible states at the same time until they are observed. It's as if the coin was both heads-up and tails-up while I have it covered with my hand. When I lift my hand up and you look at the coin, it then resolves to either heads or tails.
- Entanglement: a physical phenomenon that occurs when pairs or groups of particles are generated or interact in ways such that the quantum state of each particle cannot be described independently—instead, a quantum state may be given for the system as a whole. This is like having an entangled pair of our superposition heads/tails block and placing them on opposite sides of the coin. As soon as one of them is observed—let's say it becomes heads—we observe its entangled counterpart and find that it, too, becomes heads.

Physics is now pushing the idea of a web of quantum connections very far indeed. A significant new branch of what is sometimes called "Quantum Information Theory" has emerged, covering the ways in which information can be transmitted through a mixture of entangled states and classical information transfer. The whole subject has moved out of the realm of speculation and is now supported by increasingly elabo-rate laboratory experiments using chains of entangled pairs of particles that verify the theory in great detail. One point that emerges from this work is that information cannot be transmitted by entangled states alone because the correlations that are observed are not ones that the user can control in order to insert information; rather, the spontaneity of the par-ticles' responses is an essential part of the account. In other words, quan-tum communication always has to involve an ordinary communication channel (such as a telephone) and a quantum channel (such as entangled particles) working in tandem. So instantaneous communication (telepathy, in the sense in which it is usually conceptualized) is impossible by this means. On the other hand, empathy, in the sense of remote beings producing synchronistical related behavior, is a possibility.

Experiments have demonstrated entangled behavior many times and shown that the entangled particles "sync up" their quantum states instantly, no matter how far apart they are. That's a little surprising because the particles are exchanging information faster than the speed of light, and nothing is supposed to go faster than the speed of light. This is the basis of quantum information, coping with uncertainty at the nanometer scale.

However, when it comes to the role of entanglement in ordinary life, outside the laboratory, the situation starts to look a lot less clear. Let me put the skeptical position first. If the entanglement that is present everywhere is actually to make a difference, then the systems and organisms of the natural world need to use it in some way. The discussions of quantum information theory assume that one can prepare a pair of entangled particles, put them in two boxes, and hand one to each of two observers who take them away for later communication. But what sort of "box" does a natural organism have that can preserve a quantum state in pristine condition? The laboratory experiments using photons cannot be a precise replica of what happens in a living organism: the only known way to "store" a photon in a living system is to absorb it into the electromagnetic structure of a molecule, which is such a turbulent system that the details of the state would rapidly be lost. The only known "box" is the microtubule, studied by Stuart Hameroff, which we will describe shortly.

3.5 "Give Me Back My Hometown": Country song and de-coherence

What makes a great country song? It tells a story. It draws a line. It has a twang you can feel down to the soles of your feet. Some get mad, some get weepy, and some just get you down the road. For Eric Church's "Give Me Back My Hometown," the story takes on a more sinister tone. It opens at a funeral and travels back in time as a murder mystery involving townies who come and take what isn't theirs. It leaves a lot of questions unanswered just like a "mystery story." What can we learn from the 2015 Grammy nominee about entanglement?

To help us understand the problems that weigh against entanglement being effective in living organisms, we shall describe the way in which almost all particles are affected by a phenomenon, heavily researched over the last 20 years, called de-coherence. This is concerned with a "hidden property" of particles, namely phase. This is easy to understand in the case of a wave on water travelling past a buoy, when the buoy moves regularly up and down (with an additional regular oscillation in the direction of the wave). Here the phase of the wave at this place and at a given time is the point that the buoy is currently at in its cycle. All particles are thought of as associated with a similar wave-aspect and they carry a phase, but in general this is behind D'Espagnat's veil: there is no "buoy" that can reveal it and it is deduced only indirectly, through phenomena (in particular, interference) that are analogous to those shown by waves.

Eric Church's "Give Me Back My Hometown" video doesn't boast an easy-to-follow or linear narrative, but it's compelling just the same. The video, which includes a priest presiding over a roadside funeral, left viewers with more questions than answers.

In a small town funeral, a priest, a mafia enforcer, and several victims gather around the grave of the town's lady mafia boss, and hear her praised while the hero who killed her watches from a distance. In the lyrics, Eric Church sings about his lost love. In the video, he builds it up. Everything is lost to corruption. He did it all to save them, his girlfriend included, and they don't even have enough courage to thank him—they're too busy cowering to break free now that they have the chance. *Is there hope for the hometown he tried to save?*

This wistful, nostalgic ballad finds Church revisiting his hometown as he tries to get over what has haunted him. He sings at the track's opening:

> Damn, I used to love this view
> Sit here and drink a few
> Main Street in high school
> Up on Friday night

The emotions and scars of small town life are effectively captured and displayed. "Give Me Back My Hometown" is aimed at an ex-lover, a girl from high school who still haunts Church when he returns home. What's important is the way the star connects with his fans and the country audience as a whole with this familiar story.

> I can hear them goin' crazy
> And up here so am I
> Thinkin' about you sittin' there sayin' I hate this, I hate it
> If you couldn't stand livin' here why'd you take it, take it,

Church sings to close a first verse that describes the atmosphere at a high school football game. High school football, young love, and hometown pride are themes more country fans will be able to relate to. The song is very trope-like, residing deeply within the well-worn grooves of the often called-upon American music theme of the forgotten hometown and heartland decay. The reason this small town theme works so often is because it resonates in a fairly universal manner, especially among country music fans. Church steers the song right down country Main Street, but doesn't sacrifice lyrical integrity. His images are sharp and colorful, with each word working to add some subtle detail to a picture that's forming in one's head.

> All the colors of my youth
> The red, the green, the hope, the truth
> Are beatin' me black and blue cause you're in every scene,

Here, Church sings to begin an effective second verse. The most intriguing line is

> If you couldn't stand livin' here why'd you take it, take it.

"Take what?" one wonders. Church could be describing a very specific set of circumstances, or he could be allowing one into his subconscious. This extra depth is unique to Church's songwriting. He's a rare artist who can slide lines like that into a composition without turning the entire production into one big swing and miss. The story is relatable, and the lyric is colorful and easy to sing along with:

> Every made memory
> Every picture, every broken dream
> Yeah everything, everything, everything
> Give me back my hometown

"To be at the place that you grow up at that is your home, and the person that left you there took that from you, there's nothing lonelier than that," Church (2015) said in an interview. "So, I love that dichotomy of 'Give Me Back My Hometown,' when the person's in it, they're standing in it, and that appealed to me as a songwriter."

The mid-tempo, building track, from the lyrics, seems to strike out at someone who has skipped town, but her ghost remains and Church can't walk through town without being haunted by the memory of everywhere they used to go, like the Pizza Hut. The key phrase is:

> I used to love this view.

We can certainly add a key ingredient to "Give Me Back My Hometown": hope. No matter what has happened at the place that you grow up, remember the following:

- First, you have to trust that you can make a difference.
- Second, you have to believe that hope keeps love alive in spite of loss. Even in spite of failure.
- Third, understanding failure is critical to today's success.

Ask ourselves: do we love enough to hope, converting failure into success?

"Give Me Back My Hometown" reveals the relationship between country songs and de-coherence. De-coherence theory is about the way that the environment interacts with entangled particles so as to affect the relation of their phases. It turns out that the nature of the correlations between measurements on entangled particles is completely dependent on this phase relation. But the phases are exquisitely sensitive to perturbations by the environment, and so the influences of this can completely scramble the correlations produced by entanglement. All that is required for this to happen is that the particle states involved in the entanglement are sufficiently different for them to interact with a perturbation in different ways. If we are considering an entanglement involving the position of a large body or even a large molecule, then the slightest perturbing factor will produce enormous effects on the phase, leading to very rapid de-coherence indeed.

The de-coherence of quantum objects is a critical issue in quantum science and technology. It is generally believed that stronger noise causes faster de-coherence. Strikingly, recent theoretical work suggests that under certain conditions, the opposite is true for spins in quantum baths. An experimental observation of an anomalous de-coherence effect for the electron spin-1 of a nitrogen-vacancy center in high-purity diamond at room temperature has been reported. It is demonstrated that, under dynamical decoupling, the double-transition can have longer coherence time than the single-transition even though the former couples to the nuclear spin bath twice as strongly as the latter does. The excellent agreement between the experimental and theoretical results confirms the controllability of the weakly coupled nuclear spins in the bath, which is useful in quantum information processing and quantum metrology.

3.6   *The Bridge of Magpies: A Thanksgiving reflection about quantum entanglement*

Quantum entanglement phenomena, as exemplified in any form of the Einstein-Podolsky-Rosen (EPR) experiment, can be wholly explained by general relativity, if certain premises are adopted which may not be popular but which contradict no scientific observations to date. Those premises are that space-time is an entity akin to particles themselves (and that relativity theory describes an actual geometry of that entity), and that the specific properties of particles which are subject to entanglement are fully caused by normal mass-less boson interactions between a particle at the instant it forms and the instant it de-coheres. Given those two premises (and the uncontroversial premise that the theory of relativity is true), it is theoretically possible to deductively predict all entanglement phenomena including the results of every EPR experiment, without recourse to any special theory of quantum mechanics.

Thanksgiving is a traditional holiday for family reunion. As my family is gathering around the tables with friends, I remembered a Chinese legend about family reunion which could be helped by the progress of emerging technologies.

We know that the Milky Way is the galaxy that contains our solar system. In Navajo myth, Coyote created the Milky Way by throwing a bag up into the air. In China, this object is the river (Silver River) that separates wife and husband, the weaver girl (Zhinu) and cowherd (Niulang) who represent the stars Altair and Vega, and is bridged once a year by magpies. The weaver girl (Zhinu) and cowherd (Niulang) are reunited by a bridge of magpies on the "night of sevens," the seventh day of the seventh lunar month.

How to build the "The Bridge of Magpies"? Today, a magnetic "wormhole" that connects two regions of space can be created. A wormhole or Einstein-Rosen Bridge is a hypothetical topological feature that would fundamentally be a shortcut connecting two separate points in space-time. A wormhole is a theoretical passage through space-time that could create shortcuts for long journeys across the universe. A magnetic "wormhole" device can transmit the magnetic field from one point in space to another point, through a path that is magnetically invisible. From a magnetic point of view, this device acts like a wormhole, as if the magnetic field was transferred through an extra special dimension.

Such a magnetic "wormhole" device could cloak the passage of electromagnetic (EM) waves through an object. This wormhole could also enable objects to be hidden from external EM observation, thus the weaver girl (Zhinu) and cowherd (Niulang) can be stealthy on their trip to reunion. Figure 3.5 shows a wormhole passage through a hypothetical spacecraft with a "negative energy" induction ring.

From this theory, it is possible to create devices for practical applications, such as magnetic resonance imaging (MRI) scans. The book titled

Figure 3.5 Wormhole passage through a hypothetical spacecraft with a "negative energy" induction ring. (Courtesy of National Aeronautics and Space Administration.)

What Every Engineer Should Know About Risk Engineering and Management presented a decision tree for selecting MRI design:

- Conventional design versus innovative design

For the innovative design, MRI scanning devices emit very strong EM waves to scan bodies and body parts. The results, however, are highly magnetic fields capable of being obstructed by nearby metals and other magnetic objects, which requires mitigating risk and uncertainty during manufacturing, transportation, and application.

With this technology of a magnetic "wormhole" device, MRI scans could be used concurrently during surgical operations, as metallic surgical tools would not disturb the surrounding magnetic fields needed for the scans.

Other applications include optical computers, optical cables, and three-dimensional video displays. Metamaterials can be created for microwave and optical frequencies to allow designs for invisible cables.

The concept has the potential to be used in many practical situations in which EM frequencies would need to be transferred without distorting background EM fields.

Recent discoveries have reopened the possibility that all quantum phenomena (from wave-particle duality and probabilistic properties) might be explicable as the macroscopic effects of entirely classical systems. At the very least, the inability to reconcile quantum mechanics and relativity, despite nearly a century of trying, entails we must be willing to entertain theoretical possibilities that we presently reject. It is evident something we believe about physics must be wrong. Therefore, proposals that involve premises presently not accepted by physics cannot be ruled out merely because of that fact. This is especially the case for premises that are not widely accepted but which have not in fact been scientifically demonstrated to be false. Accepting the latter kind of premise does not entail concluding any scientific findings were "wrong," only that we were wrong to treat assumptions as if they were demonstrated facts, when those assumptions had never been scientifically demonstrated to begin with. Such an assumption is the rejection of an objective space-time.

Even more acceptable are premises no one has ever affirmed or denied because they simply hadn't been considered. Those particle properties are caused to be what they are by interactions with mass-less bosons at the moment of a particle's formation is such a premise (which bears certain similarities to but is not the same proposal as Bohm's pilot-wave theory). If these two premises are accepted, then it can be shown that relativity theory alone would entail the prediction that quantum entanglement will be observed in any given universe subject to all three conditions (those two premises, and the truth of relativity theory). In such a case, the results of EPR experiments would not be surprising, even to someone who had never even heard of quantum mechanics.

Whenever particles form (whether photons or electrons or anything else), and sometimes when they interact in certain ways, they always form or result in pairs, with certain opposite properties (such as the quantum property of spin), and for some period of time remain "entangled," such that what happens to one of the pair seems to affect its partner, in certain ramified ways. In an EPR experiment this phenomenon is tested.

In one common class of EPR experiment (which has many variations), particle "spin" is being measured. An entangled particle pair is created at an emitter and each is sent off in a different direction toward different detectors. It has been independently established that the two particles of an entangled pair always have opposite spin (one will be "right-handed" and the other "left-handed"). At each detector is a filter that blocks particles of a certain spin (such as "left-handed") but allows others through (such as "right-handed"), and the filter's ability to block or allow one kind of spin varies along a continuum as the filter is rotated (such that at most positions it allows a predictable percentage of each, and only when directly vertical or horizontal does it block all of one and allow all of the other).

What has been found in this class of EPR experiment is that the frequency of particles having a certain spin changes not merely relative to the rotated position of the filter at the detector, but relative to the rotated position of both filters together. For example, only when both filters are exactly 90 degrees out of alignment with each other do they each block all of one spin and none of the other. Rotating one filter changes this alignment, which changes the frequency of particles getting through, in a mathematically predictable way, as if somehow that filter magically "knows" at what angle the other filter has been rotated.

Any theory that can explain this will explain all entanglement phenomena because they all reduce to the same basic elements: some aspect of the physical arrangement of the system where a particle is "detected" (i.e., interacts with that physical system, which system may be our instrument, or any natural system standing in for it) affects what properties the detected particle exhibits as if it "knows" what the physical arrangement is of the other system where the other particle (the first particle's entangled partner) is "detected," even if that is thousands of miles away (or even, theoretically, infinitely distant) and even if that other system was configured only an instant before the partner particle collided with it (in fact, even if, at the moment of the particle's being formed, the other detector didn't exist), so not only does information appear to travel instantly through space, but even backward in time.

This has been demonstrated in EPR experiments where one filter is rotated after the particles have been generated and are already en route. The outcome always adheres to the arrangement of both detectors' filters relative to each other at the moment of detection, even though that arrangement didn't exist when the particles formed. This would entail either that the particles did not have any definite spin when created and information is somehow traveling instantaneously across the universe the moment one particle's spin is determined, or that somehow information is traveling backward in time from the detectors to the emitter (since whether a particle will have left-handed or right-handed spin is determined by the future state of the filters, even though, in any classical interpretation, a particle is supposed to have some spin or other the moment it is created). All varieties of quantum entanglement are iterations of this same general phenomenon: the properties that a particle appears to have upon detection are determined by the relative configuration of physical systems interacting with those particles that are far too distant from each other in time and space for any light-speed signal to have been exchanged between them.

According to pilot-wave theory, quantum phenomena are explained by "pilot waves" traveling "back to the future today" (and thus faster than light) to pre-fix the properties (such as position, spin, or momentum) of quantum particles. Though still entertained as a possibility in the physics community, it is generally rejected as untenable because it requires two implausible assumptions: non-locality (that time reversed/faster-than-light communication exists) and perfect efficiency (by requiring that the pilot signal never decays or becomes disordered despite being propagated over vast, even infinite distances).

According to established relativity theory, an object moving at a relative velocity will contract both in length and in the passage of time, in each case to a degree that is relative to the observer. This contraction reaches maximum at the speed of light, such that an object traveling at the speed of light will appear contracted to zero length and as experiencing the passage of zero time. This is not merely apparent, because in actual fact the object itself will experience the passage of zero time. If it were possible to put an astronaut in a ship traveling at the speed of light (this is impossible, but for reasons not relevant here), that astronaut would also experience no passage of time but, from his or her perspective, would have traveled instantly from source to destination, not having aged, even as his or her partner makes the trip more slowly and is considerably older upon arriving at the same destination (experiments with atomic clocks in aircraft and satellites have confirmed this principle more than amply).

If space time is an object, then it, too, experiences relative contraction. According to relativity theory, there is no objective difference between whether object A is traveling past a stationary object B or object B is traveling past a stationary object A. In fact, the issue of which is at rest (A or B) is entirely relative to the observer and has no objective status. Accordingly, if HP1 is true, then whenever we are in motion (and are not accelerating) we are at rest and space time itself is in motion. It is not only that we are in motion relative to other objects in space, but that even if there were no other objects in space, we would still be observing a space time that is moving relative to us (and anyone resting in that space time would instead see us moving relative to space time and space time itself as at rest). Accordingly, space time would contract following the same principle as objects in motion.

All mass-less bosons travel at the speed of light. The photon, actually being light, is the obvious example. But if H1 is true, then a photon experiences the passage of zero time from the moment it is created to the moment it is destroyed, and all of space time in the direction of its path will be contracted to zero length. At the moment that a photon forms, in other words, the distance between the emitter and the detector in an EPR experiment is precisely zero, and the time of travel between them is likewise zero. The distance is not zero between the emitter and detector at the moment of emission, but zero between the emitter and the future detector. In other words, the distance between the emitter and the point in space time where the photon will collide with the detector is zero. This can be demonstrated non-controversially with the non-Euclidean geometry regularly employed in relativity calculations. The whole universe is in effect contracted to zero length between the emitter and that future detector—just as any other object at rest (relative to the photon) would be.

This means that when a photon pair forms, these two photons are not "distant" from the two detectors in an EPR experiment, in either space or time, but completely adjacent to them—in fact, occupying exactly the same point in space. The physical system of the emitter and future detectors is collapsed to a distance between them all of exactly zero, and communication between them is literally instantaneous—because the passage of time for any signal sent between them is exactly zero, and the distance being traveled is exactly zero. The photon itself is therefore the only signal we need, and no additional "pilot" signal need be posited. There is also no signal decay because the distance being propagated is zero. It would not evoke any surprise at all that the spin properties of two photons at their formation would be caused to be what they are by immediately adjacent objects known to affect photon spin properties (like polarizing filters)—in fact, by objects occupying exactly the same point in space time as the photons themselves.

3.7 From Nobel Prize medicine 2015 to risk engineering of industrial products

Three scientists who developed therapies against parasitic infections have won year 2015's Nobel Prize in physiology or medicine. Most people think of parasitic diseases occurring in poor and developing countries, something they might pick up on an overseas trip. However, parasitic infections still occur in the United States, and in some cases, affect millions of people. As shown in Figure 3.6, parasitic diseases present a global health problem.

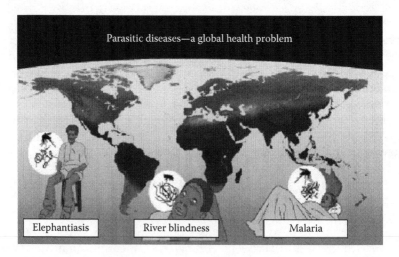

Figure 3.6 Parasitic diseases present a global health problem linked to risk engineering of industrial products. (Courtesy of www.nobelprize.org.)

A parasite is an organism that lives on or in a host and gets its food from or at the expense of its host. Parasitic diseases can be food-borne. Salmonella is the most common bacterial food-borne infection. Salmonella species are facultative intracellular parasites, capable of penetrating, surviving, and often multiplying within diverse eukaryotic cell types, including epithelial and phagocytic cells.

In the past, food would be sourced, processed, and sold locally so an outbreak would usually affect only a limited population. Today, raw materials are sourced from around the world, often at the lowest cost, and then transported long distances in complex supply chains for mass production and sale in global markets. Therefore, the severity of incidents of parasitic diseases is increasing as food products are sourced and transported globally in complex supply chains. Food producers can be affected by problems caused by small suppliers and face financial losses as a result.

For example, peanut butter is at relatively high risk because salmonella resistance to heat increases in products with low water activity and high fat content. Many types of food products contain peanut butter, including ice cream, cookies, and sweets, which extend the risk to hundreds if not thousands of final products. When an incident occurs, a large number of food producers have to recall and destroy products. Peanut butter products are often consumed by vulnerable people, including children and the elderly, increasing the risk potential of any incident.

Currently, bacterial, viral, and other parasitic diseases cause enormous suffering and millions of deaths annually, and have huge economic costs. Understanding natural immune responses and developing vaccine-induced immune responses to these and other diseases is critically important to health.

The supply chain risks and consequences experienced in peanut butter production could apply to many types of food products and producers. Similar risks are evident in other industries, such as the pharmaceutical, construction, and automotive sectors. Here, sub-standard or defective materials can cause product failures that might lead to accidents, injuries, or fatalities. As a critical element of risk engineering, mitigating risk and uncertainty is vitally important for all the industrial products.

3.8   The Imitation Game to mitigate risk and uncertainty

The Imitation Game is a 2014 American historical drama thriller film based on the biography *Alan Turing: The Enigma*, recounting newly created British intelligence agency MI6 recruited Cambridge mathematics alumnus Alan Turing to crack Nazi codes, including Enigma—which cryptanalysts had thought unbreakable.

Dr. Alan Turing introduced "The Imitation Game" in his paper, "Computing Machinery and Intelligence" (1950). The ideas proposed in this paper are a part of Turing's legacy to the fields of theoretical computer science, artificial intelligence, and machine learning. Turing proposed "The Imitation Game" as a way to answer a more precisely defined question around the level of machine learning than the question of "Can machines think?"

Dr. Turing describes the following kind of game. Suppose that we have a person, a machine, and an interrogator. The interrogator is in a room separated from the other person and the machine. The object of the game is for the interrogator to determine which of the other two the person is, and which the machine is. The interrogator knows the other person and the machine by the labels 'X' and 'Y'—but, at least at the beginning of the game, does not know which of the other person and the machine is 'X'—and at the end of the game says either "X is the person and Y is the machine" or "X is the machine and Y is the person." The interrogator is allowed to put questions to the person and the machine of the following kind: "Will X please tell me whether X plays chess?" Whichever of the machine and the other person is X must answer questions that are addressed to X. The object of the machine is to try to cause the interrogator to mistakenly conclude that the machine is the other person; the objective of the other person is to try to help the interrogator to correctly decide which of the other two the person is, and which the machine is. Dr. Turing wrote:

> Let us play the imitation game, using as witnesses a man who is good as a telepathic receiver, and a digital computer. The interrogator can ask such questions as 'What suit does the card in my right hand belong to?' The man by telepathy or clairvoyance gives the right answer 130 times out of 400 cards. The machine can only guess at random, and perhaps gets 104 right, so the interrogator makes the right identification.

There is an interesting possibility which opens here. Suppose the digital computer contains a random number generator. Then it will be natural to use this to decide what answer to give. But then the random number generator will be subject to the psychokinetic powers of the interrogator. Perhaps this psychokinesis might cause the machine to guess right more often than would be expected on a probability calculation, so that the interrogator might still be unable to make the right identification.

The psychokinetic powers can be interpreted as the special causes of statistical variations. As described by Dr. John X. Wang's book titled *What*

Every Engineer Should Know About Decision Making Under Uncertainty, a random number generator can help us simulate statistical distributions to mitigate risk and uncertainty.

3.9 Sleeping bear dunes: A legend about safety and strength

As discussed in "Safety First—Risk Engineering," safety engineering is strongly related to risk engineering and management, including assessment based on stress-strength interference. Reflecting on the legendary Sleeping Bear Sand Dune (see Figure 3.7, the most beautiful place in America), I feel that the drive for safety by maximizing strength is revealed by The Legend of Sleeping Bear.

Long ago, along the Wisconsin shoreline, a mother bear and her two cubs were driven into Lake Michigan by a raging forest fire.

"Swim," the mother bear said, "Swim as hard as you can for as long as we can, we will swim across the lake and make it to safety. If you lost sight of me over one wave, you will find me over the next."

The bears swam for many hours, but eventually the cubs tired and lagged behind.

Mother bear reached the shore and climbed to the top of a high bluff to watch and wait for her cubs.

Too tired to continue, the cubs drowned within sight of the shore.

The Great Spirit Manitou created two islands, North Manitou Island and South Manitou Island, to mark the spot where the cubs fought raging waves with their last strength.

The mother bear, who soon grew tired, saw her two cubs make it to safety in Lake Michigan as the two beautiful islands. Knowing that her

Figure 3.7 Sleeping bear dunes: a legend about safety and strength.

cubs were safe, the mother bear soon fell fast asleep, forming Sleeping Bear Sand Dune.

The Legend of Sleeping Bear is a legend about safety by overcoming risk, a legend about strength to make the journey across Lake Michigan, and a legend about love is more than raging fire and raging waves, which can be modeled by extreme value distribution in risk engineering. Based on group technology (GT), mitigating risk and uncertainty requires dedication that Mother Bear and her two cubs show for one another.

3.10   Safety first—risk engineering

Touring Michigan's Iron Mountain Iron Mine, 9 miles east of Iron Mountain on Highway U.S. 2, I saw the slogan "Safety First" (see Figure 3.8). This reminded me of safety engineering as an engineering discipline which assures that engineered systems provide acceptable levels of safety. It is strongly related to risk engineering and management. Safety engineering assures that a life-critical system behaves as needed, preventing disasters.

In 2015, explosions over the Tianjin port, the world's 10th-busiest port, were a stark reminder that it has far to go in preventing disasters—from blasts on factory floors to leaks of oil pipes and warehouse fires. Sending up huge plumes of flames, the two blasts started at a hazardous material warehouse in the eastern city of Tianjin and killed at least 114 people in one of China's worst industrial accidents in years.

According to CNBC, with about 700 tonnes of the deadly chemical sodium cyanide in the warehouse that blew up, "cyanide levels in the waters around the Tianjin port explosion site had risen to as much as

Figure 3.8 Safety first—risk engineering. (Courtesy of U.S. Department of Transportation, https://www.transportation.gov/sites/dot.gov/files/docs/safety play.jpg.)

277 times acceptable levels." With a deep-rooted business mentality that puts profits ahead of safety, the latest revelations on Ruihai International Logistics (RIL), the operator of the hazmat warehouse, suggest an unsafe business model by which safety rules can be easily bent for the convenience of the company. Being most interested in cutting costs and maximizing profits without adequate heed for safety, companies like RIL are taking chances to skimp on safety measures, and regulating agencies are unable to enforce rules.

Within days of the disaster, blatant violations of workplace safety were exposed at RIL, which was storing too much hazardous material too closely to residential homes and public infrastructure, including a light-rail station. Chinese national standards dictate that large quantities of dangerous chemicals should be stored at least 1000 m (0.6 mi) from public buildings and infrastructure. However, it was reported that the warehouse was located only some 560 m away from a nearby residential complex. Having sought to dodge safety restrictions, RIL's warehouse was storing the dangerous chemical sodium cyanide in huge amounts 70 times the limit allowed, and people are questioning whether residents in the area were suitably informed of the hazardous material.

Risk communication is a critical part of risk engineering, which puts safety first.

3.11 America on Wheels: Safe and green with the help of nanotechnology

Have you visited America on Wheels, an over-the-road transportation museum located in Allentown, Pennsylvania, in the United States? Do you know that America and the rest of the world have a long history of engineering the environmental risk of wheels? By the 1950s, traffic in California had become so heavy that smog posed a significant risk to residents. Today, China's cities have now reached a similar stage, and are increasingly following American and European regulators in imposing limits on emissions by new cars of nitrogen oxides (NOX), hydrocarbons, and fine soot particles. Other countries are also getting more concerned about global warming and pollution, and have started to require manufacturers steadily to reduce the carbon dioxide (CO_2) emissions of the vehicles they make. Having been depicted as environmental villains since the 1950s, cars and their makers may soon be able to move out of the spotlight with the help of nanotechnology.

A car is a complex combination of various components that can be converted to a greener vehicle in various ways. These may include invention of non-fossil fuel energy for these vehicles such as hydrogen fuel cells (see Figure 3.9), reduction in the consumption of fossil fuels by using

Figure 3.9 America on Wheels with fuel cells: safe and green with the help of nanotechnology. (Courtesy of Department of Energy, http://www.afdc.energy .gov/vehicles/images/fuel_cell_car.jpg.)

nanocomposites, for tire innovation as nanomaterials combined with rubber. Here are a few examples:

- Nanotechnology constitutes a certain percentage of a final product whose key functions hinge on exploiting the size-dependent phenomena of nanotechnology. Green cars are complex products which incorporate green nanotechnology in several different ways, being present in the tires, in the chassis, in the windscreen, and so on.
- Green nano-electronic manufacturing enables components of the green car, and its production through the use of sensors to reduce energy wastage and to monitor and reduce emissions.
- Electric cars using nanotechnology-enabled batteries. The battery material ($LiFePO_4$), the component electrode and the system as a whole (the battery) are all based on nanotechnology, so the final product (green car) is nano-enabled through its batteries, that is, its performance is enhanced by the use of nanotechnology.

Potential fields of improvement include increased car and truck fuel efficiency and tire durability and reduced greenhouse gas emissions and tire weight. The tire industry faces many challenges—from the on-going supply of raw materials in the face of massive predicted growth in demand for vehicles over the coming decades, to implementing innovations that will improve the sustainability of cars, for example, through the reduction of CO_2 emissions, while still delivering a high-quality product with a critical role in vehicle safety. Nanotechnology helps improve safety and sustainability by the following:

- A nanotechnology-enabled tread compound helps the tires grip the road. This tire is intended to provide higher performance than conventional tires.
- A nanoparticle enables tires to last longer, have a better grip, reduce resistance, and thus save fuel. Here, nanoparticles of rubber compound in the tire provide anchor points to attach to silica filler.

Tires thus represent a good case study through which to analyze and identify safety, technical, socio-economic, and policy issues relating to the use of green nanotechnology.

Some of the most promising nanomaterials for tires include nano-silica, organoclay, and carbon nanotubes (CNTs) used as fillers, substituting for traditional fillers like carbon black and silica.

Composites reinforced with CNTs have been shown to have dramatically improved tensile strength, tear strength, and hardness compared with more conventional materials. Silicon carbide has been used to produce tires with improved skid resistance and reduced abrasion. Nanoparticles of clay can be mixed with plastic and synthetic rubber to seal the inside of tires, creating an airtight surface and allowing the amount of rubber required to be reduced. CNT-based materials have great potential to improve the energy use of the transport sector including cars, trucks, and trains.

High-strength, ultra-light nanomaterials being used for lower-weight cellular manufacturing of vehicles are making an important contribution to higher energy and resource efficiency and substantial amounts of fuel saving, enabling America on Wheels to be safe and green.

chapter four

Design for environmental risk engineering

> The scientist and engineers who are building the future need the poets to make sense of it.
>
> **Jason Silva**
> *Media Artist*

4.1 Reflecting by Mammoth's hilly woodlands: Creating environmentally sensible products

Under a swath of Kentucky hills and hollows is a limestone labyrinth that became the heartland of a national park. The surface of Mammoth Cave National Park encompasses about 80 square miles. Today we still don't know how big the underside is. More than 365 miles of the five-level cave system have been mapped, and new caves are continually being discovered. Two layers of stone underlie Mammoth's hilly woodlands. Mammoth Cave is the heart of the south-central Kentucky karst, an integrated set of subterranean drainage basins covering more than 400 square miles. On the top of this labyrinth a biologically diverse set of ecosystems is inextricably interlinked with the underground ecosystems. This physiographic province, with Mammoth Cave National Park at its core, was declared an International Biosphere Reserve by the United Nations Educational, Scientific, and Cultural Organization's (UNESCO) Man and the Biosphere Program (MAB) in 1990.

As shown in Figure 4.1, a sandstone and shale cap, as thick as 50 feet in places, acts as an umbrella over limestone ridges. The umbrella leaks at places called sinkholes, from which surface water makes its way underground, eroding the limestone into a honeycomb of caverns. Mammoth Cave National Park encompasses 52,830 acres in south-central Kentucky and protects the diverse geological, biological, and historical features associated with the longest known cave in the world. Above the cave, the surface landscape highlights rare plants and dense forest, a diverse aquatic ecosystem in the quiet Green and Nolin Rivers.

Mammoth, a United Nations World Heritage site, still is as "grand, gloomy, and peculiar" as it was when Stephen Bishop, a young slave and early guide, described it. By a flickering lard-oil lamp Stephen Bishop found

Figure 4.1 Reflecting by Mammoth's hilly woodlands: Creating environmental sensible product. (Courtesy of National Park Service.)

and mapped some of Mammoth's passages, the world's longest Mammoth's passages. Bishop died in 1857. His grave, like his life, is part of Mammoth; it lies in the Old Guide's Cemetery near the entrance, gate to Bishop's legendary story. Mammoth Cave National Park was established in 1941 to protect the unparalleled underground labyrinth of caves, the rolling hilly country above, the heart of the south-central Kentucky karst, and the quiet Green River valley. Since then, ongoing research and exploration have shown the park to be far more complex than ever imagined, hosting a broad diversity of species living in specialized and interconnected ecosystems including green plants and vibrant animals. Today we are on a wildflower walk with a naturalist, enjoying the great diversity of flora in the national park. Mammoth Cave National Park supports more than 1300 species in about 50,000 acres.

By a flickering lard-oil lamp Stephen Bishop found and mapped some of Mammoth's passages, the world's longest Mammoth's passages. Mammoth does not glamorize the underworld with garish lighting. Here we are deep in the Earth. And nowhere else can we get a better lesson in the totality of darkness and the miracle of light. On a tour a ranger gathers everyone and, after a warning, switches off the lights. The darkness is sudden, absolute. Then the ranger lights a match and the tiny dot of light magically spreads, illuminating a circle of astonished faces. Still, many natural resources in Mammoth Cave National Park are subjected to unfavorable influences from a variety of sources, for example, air and water pollution, industrial development building up electronic waste (E-waste), and excessive visitation. Left unchecked, the very existence of many natural communities can be threatened. To help prevent the loss or impairment of such communities in the National Park System, the Natural Resource Inventory and Monitoring (I&M) Program was established. The principal functions of the I&M Program at Mammoth Cave National Park are the gathering of information about the resources and the development of techniques for monitoring the ecological communities. Ultimately, the inventory and monitoring of natural resources are integrated with park planning, operation and maintenance, visitor protection, and interpretation to establish the preservation and protection of natural resources as an integral part of park management and improve the stewardship of natural resources. The detection of changes and the quantification of trends in the conditions of natural resources are imperative for the identification of links between changes in resource conditions and the causes of changes and for the elimination or mitigation of such causes. Inventory and monitoring datasets lead to specific management actions, and then track the effectiveness of those actions. If results of resource management actions are not as anticipated, then adjustments can be made to the prescription. This is an adaptive management process toward creating environmental sensible products and mitigating E-waste.

4.2   Back to Future Green—sustaining snow for a future white Christmas

Our mission of Back to Future Green—redefining green electronics would mitigate the impacts of climate change on Santa Claus, his reindeer, and the elves.

What would it actually be like for Santa Claus, the elves, and the reindeer if the North Pole were melting?

The ice at the North Pole isn't very thick and over the last 30 years or so it has been getting a lot thinner; it is shrinking. The Greenland and Antarctic ice sheets have decreased in mass (see Figure 4.2).

The ice used to cover millions of square kilometers but now there is much less than that because of climate changes.

The littlest reindeer, Vixen, is facing great challenge to save Santa's village and workshop while the ice at the North Pole slowly melts away.

The ice at the North Pole is cracking up.

- What is going to happen to Santa and the elves? And more importantly,
- What is going to happen to the kids at Christmas?
- How are we going to sustain the snow for a white Christmas?
- What are the consequences of climate change?
- What can we do about it?
- What can our family do for climate change?

Figure 4.2 The Greenland and Antarctic ice sheets have decreased in mass. (Courtesy of National Aeronautics and Space Administration.)

These questions came as world leaders sat down in Paris to discuss a new international agreement in climate change action at the United Nations Climate Change Conference in Paris, running from November 30 to December 12, 2015.

A critical part of the answer would come from "Green Electronics Manufacturing: Creating Environmental Sensible Products," thus we can be Back to Future Green—sustaining snow for a future white Christmas, which is entangled with today's climate change. Certainly there is a constant interplay between the coherence which each system receives from the greater ones in which it is contained, and the processes of de-coherence which make it behave, in relation to its environment, as if it were a classical system. Thus, entanglement within a specific quantum state, having a function in the organism and in the greater whole, could be maintained by a top-down influence. Entanglement may be an explanation of the major paranormal experiences that many of us have encountered, but we will only arrive at a justification of this by a theoretical and experiential investigation of the cosmological level.

4.3   Eco-cruise over the river—green and blue

Reading poems on a blue heron eco-cruise,
I feel that I have known the poets before
over the great river—green and blue.

It may be that we always find our Home Sweet Home
on our journey,
when we are travelling away, far away,
so when we come back,
we find our home again—sweet and warm.
To create our own story in any of the great rivers takes real
   courage.
Few will have the greatness to bend a historic river;
However, we can change a small portion of the river flowing
   with love.
We were convinced we cannot cross the same river again.
Now we know nothing is more untrue.
We know we are coming back over and over again,
seeking the memories flowing with river.
"Let the beauty of what you love be what you do."
While we are living in eternity,
we also only have this moment
precious present moment,
flowing away from us like the river.
Of all the paths you take in life,
make sure a few of them are eco-cruises over the river—green
   and blue.

As shown in Figure 4.3, the Mississippi River is perhaps the largest and most complex floodplain river ecosystem in the Northern Hemisphere. A system is a set of connected things or parts which link together to make

Figure 4.3 The Mississippi River is perhaps the largest and most complex floodplain river ecosystem in the Northern Hemisphere. (Courtesy of National Park Service.)

the system work. Systems can be natural or artificial (manmade). Systems have inputs (things going into a system), processes (things going on within the systems), and outputs (things coming out of the system). A cake factory can be seen as a system. The raw materials (flour, sugar, etc.) are the inputs. Mixing is a process. Cakes are an output.

An ecosystem is a community of living things (e.g., plants and animals—biotic), plus the non-living things (e.g., climate, relief and soil—abiotic) they need. The parts of an ecosystem are linked together. For example, in the tropical rainforest ecosystem, rain is an input, which makes the trees grow. Evaporation is an output. There are many different ecosystems around the world. They all have unique features. Ecosystems exist at a variety of scales. Ecosystems can be small (micro). A pond is an example of a small-scale ecosystem. Medium scale ecosystems, like forests, are called meso. There are also large-scale ecosystems, for example, the tropical rainforest. Very large ecosystems are known as biomes. An ecosystem consists of the biological community that occurs in some locale, and the physical and chemical factors that make up its non-living or abiotic environment. There are many examples of ecosystems—a pond, a forest, an estuary, a grassland. The boundaries are not fixed in any objective way, although sometimes they seem obvious, as with the shoreline of a small pond. Usually the boundaries of an ecosystem are chosen for practical reasons having to do with the goals of the particular study.

The study of ecosystems mainly consists of the study of certain processes that link the living, or biotic, components to the non-living, or abiotic, components. Energy transformations and biogeochemical cycling are the main processes that comprise the field of ecosystem ecology. Sunlight is the main source of energy. This allows the plant to convert the energy into food by photosynthesis. Plants that convert energy by photosynthesis are known as producers. This allows the plants to provide food for some animals, birds, and fish. These are called herbivores. The other animals eat the animals that have eaten the plants. These are carnivores. This process is called the food chain.

An ecosystem has a series of stores and flows. In the forest ecosystem, energy and matter are stored in the wood and the leaves. There is a flow of nutrients from the soil to the leaves. These are part of cycles such as the nutrient cycle and the water cycle. Living things in the ecosystem are linked together by the flows of energy and matter as things eat each other. These links can be shown as food chains.

Ecology generally is defined as the interactions of organisms with one another and with the environment in which they occur. We can study ecology at the level of the individual, the population, the community, and the ecosystem.

Studies of individuals are concerned mostly about physiology, reproduction, development, or behavior, and studies of populations usually focus on the habitat and resource needs of individual species, their group behaviors, population growth, and what limits their abundance or causes extinction. Studies of communities examine how populations of many species interact with one another, such as predators and their prey, or competitors that share common needs or resources.

In ecosystem ecology we put all of this together and, insofar as we can, we try to understand how the system operates as a whole. This means that, rather than worrying mainly about particular species, we try to focus on major functional aspects of the system. These functional aspects include such things as the amount of energy that is produced by photosynthesis, how energy or materials flow along the many steps in a food chain, or what controls the rate of decomposition of materials or the rate at which nutrients are recycled in the system.

Measures of ecosystem function include productivity and decomposition. Ecosystem function is an important component of the health of ecosystems, along with biodiversity. Ecosystem function and biodiversity are often linked. Changes in ecosystem function are often a precursor to loss of species or changes in the composition of species in an ecosystem. In summary,

- An ecosystem consists of the biological community of a place, and the physical and chemical factors making up the abiotic environment. Ecology looks at energy transformations and biogeochemical cycling within ecosystems. Ecology is often defined as the interactions of organisms with each other and with the environment in which they occur.
- Ecosystems are made up of abiotic (non-living) and biotic components. The basic components are important to nearly all types of ecosystems.
- Energy is continually input into an ecosystem in the form of light energy, and some energy is lost with each transfer to a higher trophic level. Nutrients, on the other hand, are recycled within an ecosystem, and their supply normally limits biological activity.
- Energy is moved through an ecosystem via a food web, which is made up of interlocking food chains. Energy is first captured by photosynthesis (primary production). The amount of primary production determines the amount of energy available to higher trophic levels.
- The study of how chemical elements cycle through an ecosystem is termed biogeochemistry. A biogeochemical cycle can be expressed as a set of stores and transfers.
- A biome is a major vegetation type extending over a large area. Biome distributions are determined largely by temperature and precipitation patterns on the Earth's surface.

4.4 A whole "poetic" dynamic of the universe: Can we travel "Back to the Future Green" with quantum system engineering?

Can we travel "Back to the Future Green"? Closed time-like curves are among the most controversial features of modern physics. As legitimate solutions to Einstein's field equations, they allow for time travel, which instinctively seems paradoxical because engineers can build the time machine, and then travel back to change the original time machine. So which time machine have the engineers built?

However, in the quantum regime, the paradox like this can be resolved leaving closed time-like curves consistent with relativity. The study of these systems therefore provides valuable insight into non-linearity and the emergence of causal structures in quantum mechanics—essential for any formulation of a quantum theory of gravity. Engineers could experimentally simulate the non-linear behavior of a qubit interacting unitarily with an older version of itself, addressing some of the fascinating effects that arise in systems traversing a closed time-like curve, by which we can travel Back to the Future Green.

We can build a quantum system reproducing the behavior of a photon passing through a closed time-like curve and interacting with its older self. In other words, engineers can thus simulate a time machine quantum mechanically. Although no closed time-like curves have been discovered to date, quantum simulation nonetheless enables engineers to study these curves' unique properties and behavior.

The quantum system consists of a photon interacting with an older version of itself. That's equivalent to a single photon interacting with another trapped in a closed time-like curve. That turns out to be straightforward to simulate using a pair of entangled photons. These are photon pairs created from a single photon and so therefore share the same existence in the form of a wave function. As shown in Figure 4.4, Sandia National Laboratories leverages quantum mechanics to enable exquisite metrology devices, such as inertial sensors and frequency standards that go beyond the capabilities of conventional methods.

Sending these photons through an optical circuit gives them arbitrary polarization states and then allows them to interfere when they hit a partially polarizing beam splitter. By carefully setting the experimental parameters, this entangled system can simulate the behavior of a photon interacting with an older version of itself. The result of this interaction can be determined by detecting the pattern of photons that emerges from the beam splitter.

The system can distinguish between quantum states that are prepared in seemingly identical ways, something that is otherwise not

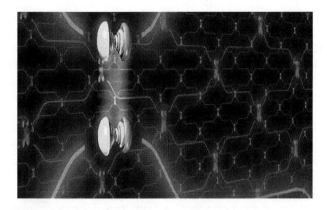

Figure 4.4 Quantum systems: Sandia leverages quantum mechanics to enable exquisite metrology devices. (Courtesy of Sandia National Laboratories.)

possible. Also, the time machine simulator can tell apart quantum states that are ordinarily impossible to distinguish. Here, quantum inputs can change the output in a non-linear way but only for some experimental setups. In other words, these inputs can control the way the experiment twists causality, which is an interesting avenue for exploring just how far it is possible to distort cause and effect, providing insights into the role of causal structures and non-linearity in our trips of Back to the Future Green. Could today really be entangled with tomorrow?

So, to summarize the skeptic's case: if we consider two particles in different places, then their states will in general be entangled. But, with the exception of the particular behavior exemplified by the polarization of photons in the laboratory, the way in which the particles are entangled, and hence the nature of any correlation between them, will be completely random, so that in practice their responses will be independent. If we consider, instead of single particles, larger systems of many particles, then the situation becomes even worse because of their greater interaction with the environment. Thus, entanglement can have no effect outside the laboratory. Historically, we have constantly found nature to surpass our own ingenuity in evolving its own subtle ways of implementing effects which we have to implement by brute force. If there is circumstantial evidence for entanglement playing a role in organisms, then there is a case for searching biological systems to discover how they might do it, even when we cannot imagine this in advance. So let us move on to discuss some areas where more solid evidence for the role of entanglement in living systems might be found, moving on from the skeptic's position to what might be called the liberal position.

This position proposes that entanglement—or at least something very like it—may play a role within an organism, as part of its internal communication and control system. In this context, Hameroff has drawn attention to the possible role of microtubules: tubes forming a "micro-skeleton" inside each living cell, made of a regular arrangement of pro-tein molecules. Because of their small size, and the way they are shielded by the structure of the surrounding water, these tubes could support internal vibrations whose states were well protected from de-coherence by the environment. Microtubules might thus form a good "box" for stor-ing quantum states. For this to be effective, however, the microtubules need to communicate with each other by conventional means: both in order to set up a state with a known entanglement (recall the need for a classical communication system alongside the quantum one) and also to keep refreshing the entanglement as de-coherence penetrates the tubules and randomizes the correlations. Hameroff, in collaboration with Roger Penrose, achieves this classical communication through a novel scheme of physics in which an aspect of gravitation, yet to be worked out in full detail, intervenes so as to realize correlated manifestations at separated microtubules, in a process that is closely linked to consciousness, with coordination happening via "gap junctions" in the microtubules. The many technical details of all this make it a very uncertain area: in 2001, for instance, Guldenagel and co-workers produced a mouse with no gap junctions but apparently normal behavior; calculations of the length of time that the entanglement can survive de-coherence are difficult and contested; Penrose's theory is still at a very speculative stage, and it is unclear how crucial it is to the coordination of the microtubules; and, at the end of all this, it is not all that clear just what the microtubules are supposed to do once they have got their act together. The liberal posi-tion leads to a lot of interesting scientific research, but in terms of the big questions of life it is not in the top league.

So the skeptical and liberal positions lead to a rather provocative situ-ation. On the one hand, entanglement seems to be consonant with some of our deepest experiences: of the connectivity of the world, of the reality of synchronicity. Yet on the other hand, it is hard to see how entanglement can act so as to actually deliver the goods. Are we somehow looking at things in the wrong way?

Many writers—including Penrose—have associated quantum pro-cesses with the mind (sometimes using in addition the word "con-sciousness"). They believe that our minds make decisions using some approximation to the formal structure of logic first described by Aristotle. But, in reality, and fortunately, this is not so: the power of our thoughts actually lies in a process that goes significantly beyond that logic, namely our ability to hold many different conceptual frameworks conjecturally together until a creative resolution emerges. And this is essentially the definition of quantum logic (the logic governing quantum systems), rather than Aristotelian logic. Is it just a coincidence that minute particles and higher mammals (let us not be too anthropocentric) share the same perverse logic? Or could it be that, as Gregory Bateson argued—with a rather careful information-theoretic definition of "mind"—all natural systems exhibit mind; and, moreover, the effect of mind is described by quantum logic?

The difficulty with linking quantum theory with these very suggestive correspondences lies in finding a way in which quantum effects can move from the microscopic, where we know they reign supreme, to the larger scale of living organisms. But could it be that this "bottom-up" approach (building the large out of smaller sub-units) inevitably leaves something out? Moreover, when we examine the skeptical and liberal approaches just outlined, it looks very much as though we are trying to extend quantum theory to the large-scale realm, while at the same time working within metaphysical assumptions about space, time, and reality that automatically exclude quantum theory from that realm. Are there alternatives to this approach?

The idea is that we regard the whole universe as a quantum system, and allow top-down influences (from the large to the small) as well as bottom-up influences. Such a perspective radically alters one's view of quantum theory: de-coherence is the loss of quantum information to the environment; however, the universe as a whole has no environment. Cosmologically, information is never lost (even, if we are to believe Hawking's recent claims, in the presence of black holes). This suggests (and there are loopholes!) that the universe remains coherent: it was, is, and always will be a pure quantum system. The non-coherence of medium scale physics—non-coherence "for all practical purposes," as John Bell used to say—is only an approximate consequence of our worm's-eye view.

When we take this viewpoint (following lines that have been explored, more conservatively, by Chris Isham and others) we find that there is a whole layer of physics revealed that is taken for granted as part of the metaphysics of laboratory physics, a layer that appears formally as the interplay of different logical structures associated with different organisms, but which we might identify subjectively as an interplay of different structures of meaning experienced by these organisms. This layer is independent of the dynamical layer investigated by laboratory physics, in the sense that, once a structure of logic/meaning emerges, then the dynamics of quantum theory operate within it without constraint, so that laboratory physics is not affected. Conversely, the outcome of this dynamic can help to shape the possible structure of logic/meaning, but it does not determine it. There is a freedom present at this level which points to a whole "poetic" dynamic of the universe.

4.5   Provide solar holiday lighting with a simple green electronics project

How to provide solar holiday lighting with a simple green electronics project? How about using the sun to power small solar and battery powered night lights, and decorations for the holiday season (see Figure 4.5)?

4.5.1   Collecting sunlight

The first part of a solar circuit is a device for collecting sunlight. To keep things simple, in my home, we're using a single nicely made small solar panel for all of these circuits. The panel that we're using for these circuits is part number PWR1241 from BG Micro, about $3 each. This is a monolithic copper indium diselenide solar panel, printed on a 60 mm square of glass and epoxy coated for toughness. On the back of the panel are two (thin) solderable terminals, with marked polarity. To prevent soldering defects, while you can solder directly to the terminals, be sure to stress-relieve the connections, for example, with a blob of epoxy over your wires. In full sunlight the panel is specified to produce 4.5 V at up to 90 mA, although 50 mA seems like a more typical figure.

4.5.2   Providing solar energy storage

To provide sufficient energy storage to power a solar circuit for extended periods of time in the dark, a rechargeable battery can deliver a fairly

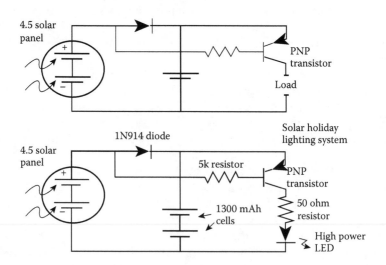

Figure 4.5 Provide solar holiday lighting with a simple green electronics project.

consistent output voltage. In the attached circuit, we use the solar panel to charge up a NiMH rechargeable battery and also LED off of the power, which will stay on when it gets dark out.

In this circuit, the solar panel charges up a 3-cell NiMH battery (3.6 V). Between the two is a "reverse blocking" diode. This one-way valve allows current to flow from the solar panel to the battery, but does not allow current to flow backward out of the battery through the solar panel. That's actually an important concern because small solar panels like these can leak up to 50 mA in the reverse direction in the dark. We're using a garden-variety 1N914 diode for reverse blocking, but there are also higher-performance diodes available that have a lower "forward voltage."

In this design we are continuously "trickle charging" up the battery when sunlight is present. For NiMH batteries and sealed lead-acid batteries (the two types that are most suitable for this sort of un-monitored circuit) it is generally safe to "trickle" charge them by feeding them current at a rate below something called "C/10." For our 1300 mAh battery cells, C/10 is 130 mA, so we should keep our charging below 130 mA; not a problem since our solar panels only source up to 90 mA.

The other thing to notice about this circuit is that it is inefficient. The LED is on all the time, whenever the battery is at least slightly charged up. That means that even while the circuit is in bright sunlight it is wasting energy by running the LED: a sizable portion of the solar panel current goes to driving the LED, not to charging the battery. Thus, we are adding a dark-detecting LED driver circuit as described below.

4.5.3 Detecting darkness

How to make a useful dark-detecting LED driver circuit? The attached circuit uses an infrared phototransistor. To add a darkness detecting capability to our solar circuit is even easier, actually, because our solar panel can directly serve as a sensor to tell when it becomes dark outside.

To perform the switching, we use a PNP transistor that is controlled by the voltage output from the solar panel. When it's sunny, the output of the panel is high, which turns off the transistor, but when it gets dark, the transistor lets current flow to the high-power LED.

Now we are using the sun to power small solar and battery powered night lights and decorations for the holiday season with a simple green electronics project.

Holiday season is just around the corner. To provide solar holiday lighting with a simple green electronics project, how do we make the solar holiday lighting twinkle just like a Twinkle Twinkle Little Star? When people start talking about blinking the high power LEDs, they often start with the heavy artillery: 555 timer chips, transistors, boost converters, microcontrollers, solid state relays, and/or dedicated LED driver chips.

While each of those does have its place, sometimes it's nice to have a simpler and much less expensive alternative.

As shown by Figure 4.6, here we describe a simple circuit for driving and blinking high-power LEDs with the blinking incandescent light bulb.

Here's the "big" yet simple idea: put the LED that you want to drive in series with a blinking light bulb. A blinking light bulb has a bimetallic strip inside that, when it gets hot enough, disconnects the circuit until it cools down. When the circuit first turns on, the light bulb and LED turn on. As the light bulb warms up, the strip bends, turning off both the bulb and LED. When the strip cools enough for the strip to snap back, the process repeats and the LED blinks. The basic circuit is scalable, if you use a high-enough power LED and light bulb—blinking bulbs like this are available in a variety of different sizes.

For our home implementation during this holiday season, we have a whopper of an LED: it's a 5 W class Luxeon K2, type LXK2-PR14-R00, with a typical radiometric power of 575 mW @ 1 A, and rated up to 1.5 A. These currently cost about $5 each in small quantities.

Our blinking bulb is a spare from a set of Christmas lights. Estimated cost: $0.15. It's designed to run with about 3 V and 100–200 mA, so that sets the scale for how hard we'll be able to drive an LED with this particular bulb. This is a fine way to run a 1 W scale LED, but if we really wanted to drive that K2 up to its full brightness, we'd need (1) a very large heat sink for the LED and (2) a bigger blinking light bulb, maybe one of the 7 W candelabra types.

For our voltage source we went tiny and used four AAA batteries, for a 6 V pack. This gives an output current in the range of between 150 and 250 mA, depending on battery freshness and bulb resistance (which is not a constant). Typical AAA alkaline batteries have a capacity near 1100 mAh, so at 200 mA and 50% duty cycle, one might expect my little battery pack to blink pleasantly for about 10 hours. Be sure to keep a multimeter handy to measure the actual currents and voltages that you're working with.

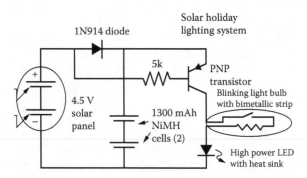

Figure 4.6 Solar holiday lighting system: Twinkle twinkle little star.

To try it out, we built up the circuit on a piece of perfboard. We soldered the back side of the LED to a 10 mil copper strip to act as a heat sink. The light bulb was not very interested in being soldered but eventually complied. After soldering all two components, it's ready to try out. So how does it work? Pretty well!

Would this circuit be inefficient? Well, yes and no. The light bulb, when it's on, basically acts as a low-ohm high-power resistor, it's no worse than any other low-ohm high-power resistor in an LED driver circuit to limit the current. Like other incandescent light bulbs, it's an efficient resistor that happens to give off a tiny fraction of its energy as visible light. That said, this isn't exactly a high-performance circuit. Working with high-powered LEDs, people may prefer to use constant-current "switching" drive circuits that operate near 85–90% efficiency, rather than the 60% that we achieve here by driving a 3.6 V LED with a 6 V source and a load resistor in between. However, this circuit is easy to construct from common materials, very inexpensive, scalable, and easy to understand. And being environmentally sensible using green energy—solar energy—the Solar Holiday Lighting System twinkles just like a Twinkle Twinkle Little Star.

4.6   Back to a sustainable future: Green computing toward lower carbon emissions

Green computing is the study and practice of environmentally sustainable computing. This can include designing, manufacturing, using, and disposing of computers, servers, and associated subsystems—such as monitors, printers, storage devices, and networking and communications systems—efficiently and effectively with minimal or no impact on the environment. The goals of green computing are

- Reduce the use of hazardous materials,
- Maximize energy efficiency during the product's lifetime, and
- Promote the recyclability or biodegradability of defunct products and factory waste.

Green computing is important for all classes of systems, ranging from handheld systems to large-scale data centers, since people these days have become more environmentally conscious, and the green trend is here to stay. Product designers have realized that everyone is going eco-crazy, whether that means going on green vacations, using green electronic products, and even having green weddings. And today, when designing any product, whether it's a computer, a couch or the latest smart phone, being environmentally friendly is almost a requirement. Of course, this goes without saying that green product demand has also increased and environmentally friendly products not only save money, but get profits flowing in.

4.6.1   *Today's challenge with carbon emissions*

An important component of creating greener computing products is lowering the amount of energy used and carbon emissions released during the entire production process, from the gathering of materials to the shipping vessels used to transport them to stores. Think about how far laptops and smart phones had to travel to get your hands—and what went into making them. Today, the majority of personal computing products are manufactured overseas, where carbon emission regulations are vastly different from country to country.

Related to green computing, the idea of "green electronics" in part refers to the materials used in the production of computers, cell phones, televisions, and dozens of other electric devices. For example, the plastic we see on all of those cable cords is often burned off in order to reach the valuable copper within, releasing dangerous chemicals into the air during the process. This can lead to the development of acid rain and further environmental damage. Let's look at the ways that this green trend has influenced product design.

4.6.2   *Green computing product design criteria*

To design a truly green computing product, it must meet some or all of these criteria:

- Be nontoxic so as not to harm the environment, people, and pets; electronics products, for example, must contain lead-free printed circuit board (PCB).
- It can be recycled or recyclable, to reduce the amount of trash in the landfills.
- It must use energy responsibly, whether that means that products use only renewable energy sources such as wind, solar, or geothermal power, or will reduce energy use, such as electronic products that go into "sleep mode" to conserve energy.
- To a certain extent, it must support environmental responsibility, such as eco-friendly practices, creating more green or local jobs, and even use fair and truthful marketing when selling their products

4.6.3   *Green materials*

Understanding the materials used for any process is essential for any project and one of the first things many designers must master is the use of materials. Unlike 20 years ago, eco-friendly materials are now more available than ever. Eco-friendly plastics, for example, which can be recycled or biodegradable, are now more widely available, but are also as tough and durable as their regular counterparts. Take the ubiquitous plastic water bottle, for example—simple to design but the material takes hundreds of years to decompose, and is quite toxic to the environment. Arizona-based Enso Bottles has developed a truly biodegradable plastic, by using an additive that helps the bottle degrade in as little as 250 days, without releasing any harmful gasses. Electronic manufacturers also use green materials for their own products. For example, LCD TVs that use carbon neutral biopaint, smart phones with bioplastic enclosures, and electronic products that feature lead-free electronics PCB boards.

4.6.4   Green product manufacture

It's not enough that your materials are eco-friendly, but the way you create your product should be as well. Consumers truly care about how a product is made, and so the construction of a product must also fit within green standards. For example, Kyocera, a Japanese firm, creates their own energy from solar power generating systems for their manufacturing plants and offices around the world. One of the problems of any manufacturing plant is not just the energy they use, but the amount of waste produced. Here, electronic waste or e-waste is a prevalent problem, this time on the side of electronic product designers. In many cases, such as with the CEH (Center for Environmental Health) in the United States, electronic design houses are encouraged to design products that are eco-friendly and safe for the environment, whether that means creating nontoxic programs or creating products that can easily be recycled.

4.6.5   Green product disposal

Aside from just waste disposal, the end-of-life disposal is just as important—what happens when a product is no longer useful and must be replaced? Previously, manufacturers just let their old products linger in landfills, but for today's environmentally conscious consumer, that simply won't do. Many manufacturers recycle their products, or donate their waste to other companies or organizations that can reuse their old materials. Electronics designers and manufacturers should, from the very beginning of the design process, create "take-back" campaigns wherein consumers are encouraged to bring their used electronics back to the manufacturer for proper disposal or better yet, recycling. Apple Computers in 2009, for example, figured out that they were emitting 9.6 million metric tons of greenhouse gases every year. So, within the next year, they re-evaluated their entire process—from designing, to manufacturing, transportation, product use, recycling, and even how they built their facilities (office, stores, etc.) and made numerous changes that drastically reduced their carbon emissions. Their biggest expenditure when it came to carbon emissions was the manufacturing process itself (45%) and so they drastically reduced this by redesigning their products to be smaller, thinner, and lighter, thus dramatically lowering their overall carbon footprint, ensuring a sustainable future, toward environmentally sensible green computing.

4.6.6   Back to the Future Green—redefining "green electronics"

Back in 1985, in the second installment of the *Back to the Future* movie trilogy, Marty McFly, played by Michael J. Fox, travels 30 years into the future to stop his future son from making a big mistake. The world he finds on October 21, 2015, is much changed from the one he left behind, and while there is no shortage of debate on the things the movie got right and wrong, we know Back to the Future Green holds the promise for a sustainable future.

Every year, hundreds of thousands of old computers and mobile phones are disposed of unsustainably. Children are exposed to a hazardous mix of toxic chemicals and poisons. The rate at which these mountains of obsolete electronic products are growing will reach crisis proportions.

The amount of electronic products discarded globally has skyrocketed recently with 20 to 50 million metric tons generated every year. If the estimated amount of e-waste generated every year were put into containers on a train, it would stretch all the way around the world. According to *U.S. News*, "A rising mountain of hazardous electronic waste is putting workers in developing countries and the environment at risk."

Electronic waste now makes up 5% of all municipal solid waste worldwide—nearly the same amount as all plastic packaging—but it is much more hazardous. The e-waste problem is growing because people are upgrading their mobile phones, computers, televisions, audio equipment, and printers more frequently than ever before. Mobile phones and computers are replaced most often.

The electronics of tomorrow hold both great technological promise and the potential for environmental peril. Things like 3-D printers, wearable, converged devices, and the cloud will continue to revolutionize how we interact with the world and with one another. Underlying that innovation are new materials, new design challenges, new supply chain concerns, and new questions about disposal and extended life. Each of these topics offers an opportunity for environmental success. And each of them must address the new consumers and regions that are becoming reliant on these devices.

Back to the Future Green, the first step on the journey is rethinking what it means to be a green electronic. Creating environmentally sensible products is implemented to manufacture clean, durable products that can be upgraded, recycled, and disposed of safely. The evolution of technology and our environmental health are inextricably linked. We must work together to address these challenges head on. Here's the high-level revelation: if we want to buy a smartphone, we should be looking at models that are free of hazardous polyvinyl chloride (PVC) plastic or brominated flame retardants (BFRs), which present environmental problems and human health concerns throughout their life cycles. Each of us has an important role in redefining "green electronics" for our journey of Back to the Future Green.

4.7 Solar roasted turkey: Have a green Thanksgiving

How to have a green Thanksgiving and add a touch of green to our Thanksgiving dinner? November is a great month for solar cooking. A solar cooker makes a great adjunct or primary for cooking the Thanksgiving meal. Solar cooking is simply harnessing the sun's energy to cook food. A solar oven uses the power of the sun to cook. In our climate, that means that the majority of days we can cook our meals outside without using additional energy, saving our precious fuel storage. Solar energy is a clean, inexpensive, abundant, renewable energy.

While roasted turkey in the solar oven is an easy way to cook our Thanksgiving bird, the weather forecast for Thanksgiving Day here in Grand Rapids, Michigan, isn't looking favorable for solar cooking. I'm afraid this year we will have to plan on roasting our turkey indoors. However, if you are lucky enough to be in a part of the world where you'll get enough sunshine, solar roasted turkey is as easy as it is delicious. When I lived in Tucson, Arizona, I was able to cook a 14-pound bird (unstuffed) in three and a half hours. Here's how it's done:

- Remove the leveling tray from the sun cooker's chamber.
- Place a rack on the bottom of it.
- Put the seasoned turkey in the sun cooker's roasting bag.
- Place a probe thermometer in the thickest part of the thigh, seal the oven bag.
- Place the turkey in the sun cooker.
- Roast until thermometer reaches 180°F. Carefully transfer the cooked turkey to a large roasting pan.
- Cut bag open, allowing cooking juices to drain into the pan.

- Transfer turkey to a cutting board and let rest for 10 minutes before carving.
- Use cooking liquid to prepare gravy.
- Enjoy!

Have a green Thanksgiving and add a touch of green to our Thanksgiving dinner. Thus, we can "Back to the Future Green" in the near future.

4.8   Saga of Singapore: Story about criticality to achieve sustainability

Time does heal all wounds when it comes to the shifting sands of Lake Michigan. The history and the mystery are part of the saga of lost mill city—Singapore—telling the criticality to "achieve sustainability to support future generations and at the same time preserve our natural resources."

Singapore, perhaps Michigan's most famous ghost town, is not one of the casualties of the four great fires (Chicago, Holland, Peshtigo, and Manistee) that ravaged the northern Midwest on October 8, 1871. Instead, it is a casualty of unsustainable economy. When its mills had eaten most of the virgin White Pine of the area, its days were numbered.

Once upon a time, Singapore was a bustling lumber town located just north of Saugatuck on the banks of Lake Michigan (see Figure 4.7). The

Figure 4.7 Singapore was a bustling lumber town located just north of Saugatuck on the banks of Lake Michigan. (Courtesy of State of Michigan, https://www.michigan.gov/documents/dnr/saugatuck_dunes_gmp_complete_322078_7.pdf.)

bustling waterfront city existed for almost half a century as a busy port and shipbuilding town that also churned out upward of 300,000 board feet of lumber a month courtesy of the three sawmills harvesting the abundant White Pine covering the surrounding land. It was boom or bust in the days of America's rapid growth.

Unfortunately, Singapore became a casualty of unsustainable economy. After the fires which swept through Chicago, Holland, and Peshtigo in late 1871, Singapore was almost completely deforested after supplying the three towns with lumber for rebuilding. The demand for timber soon depleted the Allegan County forests. Without the protective tree cover, stripped of timber and blown by a near-constant west wind off the lake, the dune between Singapore and the lake enveloped the town's streets and buried all but the tallest structure. Eventually, that too, was hidden. The winds and sands coming off Lake Michigan quickly eroded the town into ruins and within four years had completely covered it over.

The town of Singapore was vacated by 1875. Its ruins now lie buried beneath the sand dunes of the Lake Michigan shoreline at the mouth of the Kalamazoo River in Saugatuck Township, near the cities of Saugatuck and Douglas in Allegan County.

Was it really "Michigan's Pompeii"? Yes, it's not a stretch to presume beneath those dunes sits remnants of Michigan's own Pompeii. Apparently time does heal all wounds when it comes to the shifting sands of Lake Michigan. The ghosts are another story all together, one with a life of their own, telling the need to "ensure higher (and sustainable) productivity and efficiency in every line of a business over the long term."

4.9   Nano research and applications enable green additive electronics manufacturing

Green electronics manufacturing creates electronic products that have minimum adverse environmental effects throughout their life cycle, which means:

- Harmful materials are not used or generated.
- Eco-efficient and environmentally sensible manufacturing processes are adopted.
- Less power is consumed in operation and standby modes.
- Fully recyclable with no hazardous waste.

Key elements to implement the green electronics manufacturing are:

1. Sustainability: meeting the needs of society in ways that can continue indefinitely into the future.

2. Life cycle (cradle to cradle) design: ending the cradle-to-grave cycle of manufacturing by creating products that can be fully reclaimed or reused or recycled.
3. Source reduction: reducing waste and pollution by changing patterns of production and consumption.
4. Innovation: developing alternatives to technologies through TRIZ (theory of inventive problem-solving) and axiomatic design.
5. Viability: creating economic activities that are friendly to the environment.

The typical approach in electronics manufacturing has been subtractive—to remove materials to make the final product useful, for example, to cut or drill materials away, leaving only the desired substance. Subtractive electronics manufacturing generates a lot of waste (piles of scrap), which are detrimental to the environment and ecosystems.

Incorporating 3D printing and laser sintering into electronics manufacturing, additive electronics manufacturing (AEM), a branch of additive manufacturing, adds material in layers to create the desired three-dimensional devices. The desired devices can be made of materials that can be sprayed as liquids (e.g., plastics or resins) or formed from melted powder (including metals and ceramics). The devices are typically manufactured directly from a 3D CAD image. Comparing with subtractive electronics manufacturing, AEM reduces/minimizes waste generated from material removal, and thus:

- Saves material cost,
- Consumes less energy during manufacturing/production,
- Improves productivity,
- Prevents pollution/contamination of the environment.

Currently, the narrow choice of materials used in AEM remains a key limitation to more advanced systems. In nanotechnology, nanomaterials with specially tailored properties can be modified by ultra fine particle size, crystallinity, structure, or surface condition. By leveraging the modifiable properties of nanostructures, electronic devices can be created from the bottom-up by adding material one cross-sectional layer at a time, expanding the material properties, and thus applications, of AEM. Integrating multiple nano-electronic devices in the same AEM part would then allow us to move from simple to more complex printed objects such as "smart" products from phones to tablets, watches, and glasses. Recent advancements in nano-biomaterials could also further enable the printing of replacement organs and bone, leading to advances in tissue repair for wound healing. Being able to start from atoms to form the desired structure, nano manufacturing paves a way to provide sustainable green electronics manufacturing by minimizing cost due to materials removal, and thus creates environmentally sensible products that are cost-effective.

In summary, nano research and applications enable green additive electronics manufacturing, creating electronic products that have minimum adverse environmental effects throughout their life cycle. Furthermore, integrate nanotechnology and additive electronics manufacturing could facilitate cellular manufacturing of electronic products by providing more flexibility in device geometry, PCB layout/topology, and electronic materials.

4.10   We are young as long as our world is green

Looking at the photo that air pollution is shrinking scenic views, damaging plants, and degrading high elevation streams and soils in the Great Smoky Mountains (see Figure 4.8), we might remember a Chinese lyric "Memorial Night" which contains the following:

Man is young as long as mountain is green.

This reminds me of the book titled *Green Electronics Manufacturing: Creating Environmental Sensible Products*.

Figure 4.8 Air pollution is shrinking scenic views, damaging plants, and degrading high elevation streams and soils in the Great Smoky Mountains. (Courtesy of National Park Service.)

Today, while turning 110 years old, the electronics industry remains a young industry facing the challenge of creating environmentally sensible and reliable "green" electronics. By chance, electronic waste (e-waste) is also described by a mountain.

We should remember the following facts/forecast:

1. The global volume of e-waste generated is expected to reach 93.5 million tons in 2016.
2. Up to 85% of electronic products were discarded in landfills or incinerators, which can release certain toxins into the air.
3. E-waste represents 2% of America's trash in landfills, but it equals 70% of overall toxic waste.
4. The extreme amount of lead in electronics alone causes damage in the central nervous systems, peripheral nervous systems, blood, and kidneys.

This emphasizes the critical importance of lead-free electronics reliability.

Cell phones and other electronic items contain high amounts of precious metals like gold or silver.

Only 12.5% of e-waste is currently recycled.

For every 1 million cell phones that are recycled, the following can be recovered:

- 35,274 lb of copper
- 772 lb of silver
- 75 lb of gold
- 33 lb of palladium

Recycling 1 million laptops saves the energy equivalent to the annual electricity used by 3657 U.S. homes.

E-waste is still the fastest growing municipal waste stream in America, according to the U.S. Environmental Protection Agency (U.S. EPA).

A large number of e-waste is mislabeled. Instead, the e-wastes are whole electronic equipment or parts that are readily marketable for reuse or can be recycled for materials recovery.

The following resources are taken to manufacture one computer and monitor:

- 539 lb of fossil fuel
- 48 lb of chemicals
- 1.5 tons of water

Electronic items that are considered to be hazardous include, but are not limited to:

- Televisions
- Computer monitors that contain cathode ray tubes
- LCD desktop monitors
- LCD televisions
- Plasma televisions
- Portable DVD players with LCD screens

Yes, "man is young as long as mountain is green." For global risk engineering including ecosystem risk engineering, we should remember

We are young as long as our world is green.

4.11   Engineering environmentally sensible and reliable "green" electronics

Many scientists and engineers consider the birth of electronics to be November 16, 1904, when U.K. engineer Sir John Ambrose Fleming applied for a British patent for his invention: the thermionic diode.

In 1880, Thomas Edison discovered electrons boiling off of a heated filament and filed a patent on the effect, which is called the "Edison effect."

In 1904, based on the Edison effect, Sir John Ambrose Fleming invented the two-electrode radio rectifier or vacuum diode. The valve demonstrated its reliability by successfully detecting radio waves. While Fleming called his device an oscillation valve, it has also been referred to as a thermionic valve, vacuum diode, and a Fleming valve.

On Christmas Eve 1906, engineering professor Reginald Fessenden transmitted a voice and music program in Massachusetts that is picked up as far away as Virginia.

In 1907, transcontinental telephone service became possible with Lee De Forest's patent of the triode, or three-element vacuum tube, which electronically amplifies signals.

Today, while turning 110 years old, the electronics industry remains a young industry facing the challenge of creating environmentally sensible and reliable "green" electronics.

4.12   From the world's first working laser to vertical-cavity surface-emitting lasers

The book titled *Green Electronics Manufacturing: Creating Environmental Sensible Products* presented how to integrate vertical-cavity surface-emitting lasers (VCSELs) onto integrated circuits. VCSELs are semiconductor laser diodes with laser beam emission perpendicular from the top surface, contrary to conventional edge-emitting semiconductor lasers which emit from surfaces formed by cleaving the individual chip out of a wafer.

An acronym for light amplification by simulated emission of radiation, the laser is unique because of its narrow, coherent beam. The use of lasers varies from laser printers and barcode scanners to surgical treatment and fingerprint detecting, of great importance to risk engineering in numerous industries. Dr. Theodore Maiman's laser—the world's first working laser—represented a major breakthrough in the field of applied physics.

In November 14, 1967, a U.S. patent for "Ruby Laser Systems" was issued to American physicist Theodore Maiman (No. 3,353,115), for which he had applied on April 13, 1961. He had first operated a laser based on a ruby crystal on May 16, 1960.

"Ruby Laser Systems" generate high chromium content, and absorb green and blue light while emitting red light. By flashing white light into a cylinder of ruby, Dr. Maiman energized the electrons in the chromium. The energized green and blue wavelengths were absorbed and then amplified the red wavelengths until the light pulse of the ruby was amplified to high power, resulting in a laser. One of Maiman's greatest contributions to the field of laser technology was his demonstration of the ease with which lasers could be constructed, making them of great practical use.

With Dr. Maiman's successful test of the first laser, a laser boom began. By 1961, the first commercial laser hit the market. Laser technology increased as rapidly as the commercial laser industry. Fast on the heels of Maiman's laser came various laser systems including:

- Dye laser
- Helium-neon laser
- Semi-conductor laser
- Carbon-dioxide laser
- Ion laser
- Metal-vapor laser
- Excimer laser
- Free-electron laser

For his outstanding work, Dr. Maiman was nominated twice for a Nobel Prize. He was also the recipient of many other prestigious awards and group memberships.

As discussed by Dr. John X. Wang's book, VCSELs can be bonded to integrated circuits (ICs) through flip-chip bonding of optoelectronic ICs.

4.13   Green electronics manufacturing of transformers at age of extinction

According to an article in the *Wall Street Journal* (WSJ) in 2014, the movie *Transformers: Age of Extinction* presented an advanced case of metal fatigue.

"The Rules Have Changed," the movie emphasized. Here are a few questions for developing the new Transformers at Age of Extinction:

- Are we free to change the rule of fatigue life?
- Can we still apply S-N (stress vs. number of fatigue cycles) curves to assess transformers' fatigue life?
- Can we perform probabilistic design based on the elastic-plastic behavior modeling?

Based on the book titled *Green Electronics Manufacturing: Creating Environmental Sensible Products*, the rule of fatigue life is essential for a successful Green Transformer Reliability Program at the Age of Extinction, considering the extremely high vibration and high temperature environments.

Consideration of fatigue reliability during the design process can assist in the prevention of failures of structural and mechanical components subject to fluctuating loads in service. Explicit consideration of the reliability of structural and mechanical components provides the means to evaluate alternative designs and to ensure that specified risk levels are met. Probabilistic fatigue analyses may also be applied to life extension of existing structures, and for problem assessment of in-service fatigue failures.

The results of a probabilistic fatigue analysis are usually expressed as the probability of failure as a function of time. For dealing with very high reliabilities, the reliability index, β, is often used as defined by Equation 4.1:

$$\beta = \Phi^{-1}(R) = \Phi^{-1}(1 - P_f) = -\Phi^{-1}(P_f) \tag{4.1}$$

in which
   β = Relativity Index
   Φ^{-1} = the inverse standard normal cumulative distribution function
   P_f = the probability of failure
   R = Reliability

An example of the change in reliability with time due to fatigue is given in Figure 4.9. The corresponding plot of reliability index as a function of time is given in Figure 4.10. The expected or average time to failure for this example is 1265 seconds, at which time the reliability is 50% and the reliability index is zero. Also shown in Figures 4.9 and 4.10 are

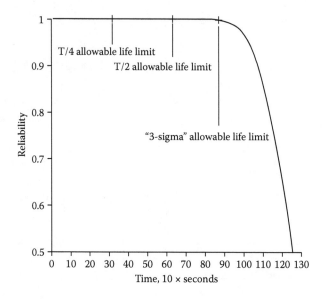

Figure 4.9 An example of the change in reliability with time due to fatigue.

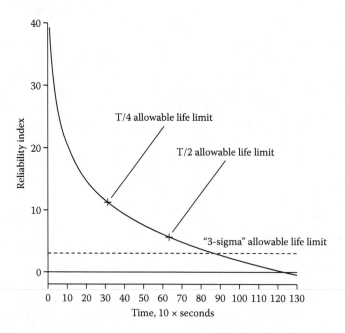

Figure 4.10 Fatigue reliability index as a function of time.

the times corresponding to the usual definition of "safe life" determined by using a scatter factor of 2 to 4. Using a scatter factor of 4 results in a safe life of 316 seconds; in comparison, a scatter factor of 2 gives a safe life of 632 seconds. Adoption of a reliability approach would allow specification of the safe life as the time at which the reliability or reliability index decreased below some minimum acceptable level. In this example, that level was chosen to be the "3-sigma" level, at which the reliability is 99.865% or the reliability index is 3.0. This results in a safe (or allowable) life of 869 seconds, a 37% increase in allowable life compared to the scatter factor of 2 life, and a 175% increase compared to the scatter factor of 4 life. The conservatism inherent in the deterministic scatter factor approach is illustrated more clearly in Figure 4.10. In this example, a scatter factor of 2 corresponds to a reliability index of 5.6, while a scatter factor of 4 corresponds to a reliability index of 11.2.

Use of a probabilistic method allows the uncertainty in fatigue life or time to failure to be more explicitly accounted for than is possible in the "scatter factor" approach. Sensitivity analysis methods are also available in some of the commercial software packages, allowing the uncertain input variables to be ranked according to their contribution to the uncertainty in the resulting life estimate. This enables a timelier and cost-effective design optimization by identifying the most important input parameters upon which resources should be concentrated to gain the greatest increase in life. This approach naturally leads to a design risk assessment and requires the development and selection of specific criteria for defining acceptable risk or probability of failure. The acceptable risk value should be specified in the appropriate program requirements, but the selection process is a program and policy issue which is beyond the scope of this guideline. Some guidance is available from the various national civil structural requirements codes that have been formulated to provide reliability indices of 3 to 5, depending on the importance of the structure and the consequences of failure.

This method is primarily useful as a design evaluation or optimization tool and cannot easily be used to evaluate the remaining life of an in-service component unless instrumentation (such as strain gages, accelerometers, or load cells) is in place to allow collection of data on the actual performance of the component. In cases where the necessary in-service data is available, this method may prove very useful for extending the life of the component or structure beyond its original predicted service life. If inspection of the component or structure by a non-destructive evaluation (NDE) method is feasible, the probabilistic fracture mechanics method described in the companion guideline is more useful for estimating damage and remaining life of in-service components than the fatigue analysis method. A measured crack length provides information about the in-service state of the structure, and the probabilistic fracture analysis may be updated following the inspection to this new information. If inspection is not feasible, the fracture mechanics method has no particular advantage over the fatigue method for in-service assessments.

The potential loss of strength in structural/mechanical components due to the cumulative damage effect of fluctuating applied loads is well known. Spectacular failures have resulted from fatigue since the beginning of the industrial revolution. The danger of fatigue in new applications, however, has not always been adequately considered in the design process and continues to be a concern to this day. A classic example is the British Comet airliner, which was one of the first aircraft to employ an aluminum-skin and pressurized fuselage. The loss of three aircraft and many lives in 1953 and 1954 occurred before the potential for fatigue in this application was understood.

Results from the theory of stochastic process and modern structural reliability methods enable the design engineer to assess the expected fatigue life of structural and mechanical components subject to randomly varying loads, and to estimate how the probability of failure of the component increases over time. The application of these data will enable optimal designs to be achieved which balance the initial costs of design and fabrication against the expected costs of repair, replacement, and/or failure.

4.14 "Gone with the Wind" always: De Zwaan as the only authentic, working Dutch windmill in the United States

Twenty years ago, on the way to the Delft University of Technology in the Netherlands, my wife Lisa and I visited Krommenie near Amsterdam, where De Zwaan windmill was first erected in 1761.

As the only authentic, working Dutch windmill in the United States, De Zwaan has experienced a unique ongoing journey of engineering design, construction/reconstruction, production (grain milling and processing), maintenance, risk mitigation/management, and wind energy utilization over 255 years.

The Persians built the first (recorded) windmills around 900 A.D. These vertical axis windmills were not very efficient at capturing the wind's power and particularly susceptible to damage during high winds. During the Middle Ages, wind turbines began to appear in Europe. These turbines resembled the 4-bladed horizontal axis windmill typically associated with Holland. The applications of windmills in Europe included water pumping, grinding grain, sawing wood, and powering tools. Like modern wind turbines, the early European systems had a yaw degree of freedom that allowed the turbine to turn into the wind to capture the most power. The use of windmills in Europe reached its height in the nineteenth century just before the onset of the Industrial Revolution. At this time, windmill designs were beginning to include some of the same features found on modern wind turbines including yaw drive systems, air foil shaped blades, and power limiting control systems.

Public and private interest in wind energy increased in the 1960s as environmental concerns about fossil fuels and pollution began to emerge. Significant wind energy research was not conducted in the United States until the Oil Crisis of the mid-1970s. During the California Wind Rush (1980–1985), thousands of wind turbines were installed in California.

Wind turbines have continued to evolve over the past 20 years and the overall cost of energy required to produce electricity from wind is now competitive with traditional fossil fuel energy sources. This reduction in wind energy cost is the result of improved aerodynamic designs, advanced materials, improved power electronics, advanced control strategies, and rigorous component testing.

Modern wind turbines are fatigue critical machines that are typically used to produce electrical power from the wind. Operational experiences with these large rotating machines indicated that their components (primarily blades and blade joints) were failing at unexpectedly high rates, which led the wind turbine community to develop fatigue analysis capabilities for wind turbines. Our ability to analyze the fatigue behavior of wind turbine components has matured to the point that the prediction of service lifetime is becoming an essential part of the design process.

Additionally, offshore wind turbine generators (OWTGs) are subjected to fatigue loads arising from several random conditions associated with the demanding environments in which these systems are designed to operate. The design of an OWTG attempts to maximize the power production by optimizing the rotor blade diameter and the rotor/nacelle control system to capture as much of the wind force as possible. However, the design of the tower structure also should be optimized, which is achieved generally by making it as light as possible. For floating OWTGs, an additional challenge is introduced in that the system on which the tower is supported must be designed with efficient consideration of the mooring response and damping effects. Thus, a satisfactory tower design is achieved when the competing interests are balanced between:

1. Maximizing the wind force captured and
2. Providing a structurally efficient design.

Subsequently, the projected power production from the OWTG can be evaluated against the cost to implement the design and to ensure the level of return appropriate for the investment.

In all cases, the tower structures specifically require due consideration of the design of fatigue critical details and the potential for crack initiation and crack progression. Consistent with any fatigue design, all sources potentially contributing to the accumulation of fatigue damage need to be identified for design. For OWTGs, the fatigue loading is dominated by the wind loads, wave loads, and operational loads, which may act alone or in combination to produce the long-term stress history used to determine the structural response. The behavior of the OWTG tower structure in response to the random loads is complex and must consider the static and dynamic response and the aerodynamic damping produced by the turbine rotor operating in the environment. Then, the damage attributed to the load cycles in the long-term stress history needs to be characterized and applied consistent with a design philosophy, such as safe life, damage tolerant, or fail-safe to estimate the fatigue performance. A reliable estimation of the fatigue life is fundamental to ensuring the design of the OWTG will function throughout the expected service life and allow the opportunity to optimize power production.

4.15 Great Smoky Mountains: Risk engineering of our grand ecosystems

The Great Smoky Mountains is the country's most popular national park. As a UNESCO World Heritage Site, the United Nations has designated the Great Smoky Mountains International Biosphere Reserve. Functioning as one of the most diverse ecosystems on Earth, the U.S. national park remains a temperate rainforest that is world-class lush.

From the design for risk engineering perspective, the ecosystem here is featured by diversity:

- A huge area of the Great Smoky Mountains National Park soars above 4500 feet to more than 6600 feet. Here, cool summers and deep-snow winters (see Figure 4.11) prevail in the spruce-fir ecosystem of the Canadian forest zone consisting of red spruce and Fraser fir.
- In contrast, vast portions of the national park sprawl below 2500 feet, where a diverse pine and forest reigns. Summer can bring sticky, even hot weather.

Between the two ecosystems, elevation and aspect yield woodlands of different types.

John Muir (1838–1914), America's most famous and influential naturalist and conservationist, said: "The mountains are fountains of men as well as of rivers, of glaciers, of fertile soil. The great poets, philosophers, prophets, able men whose thoughts and deeds have moved the world, have come down from the mountains—mountain dwellers who have grown strong there with the forest trees in Nature's workshops."

Figure 4.11 A deep-snow winter at the Great Smoky Mountains. (Courtesy of National Park Service.)

Travelling back from the Great Smoky Mountains to the Grand River, another grand ecosystem, I feel (as John Muir did), that we are all mountaineers travelling over rivers of no return.

4.16   Ebola prevention: Risk engineering of our global health system in our times

On December 10, 2014, *TIME* announced choices for Person of the Year 2014. They chose the Ebola Fighters, who risked, persisted, sacrificed, and saved precious life.

For decades, Ebola terrified rural African villages like some mythic monster. Every few years, the disease rose to demand a human sacrifice and then returned to its cave. Like a Hollywood horror, Ebola reached the rest of world only in nightmare form. Ebola virus disease causes eyes to bleed and organs to dissolve. Worst of all, Ebola makes doctors despair because they have no cure. The Ebola virus, pictured in Figure 4.12 from a special type of microscope, is the agent that causes Ebola Hemorrhagic Fever (EHF).

The Ebola crisis is disturbing and alarming in many ways. Let's apply risk priority number (RPN) to assess Ebola from the risk engineering perspective.

$$RPN = Occurrence\ Probability \times Severity \times Escape$$

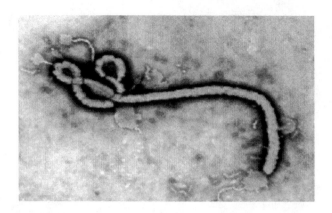

Figure 4.12 The Ebola virus, pictured above in the image from a special type of microscope, is the agent that causes Ebola Hemorrhagic Fever (EHF). (Courtesy of Centers for Disease Control and Prevention, United States.)

4.16.1   Occurrence probability

2014 is the year an outbreak turned into an epidemic, powered by the rapid spread globally. The gloves, mask, and other gear used for infection control are undoubtedly very protective. However, when used in the real world, as opposed to in the laboratory, they cannot possibly be "completely" protective—a fact that should have been suspected earlier, and has been proven now in transmissions to health care workers in both Texas and in Madrid. Each time a health worker cares for an Ebola patient, he or she wears and then removes protective gear. Only a very small probability of an exposure exists. However, over many repetitions, that probability gets amplified. If we perform a certain activity once that has a very low probability of a very negative consequence, our risks are low. But if we repeat that activity many times, the laws of probability, or more precisely, a statistical distribution called the "binomial distribution" will eventually catch up with us.

4.16.2   Severity of consequence

Beyond the crowded slums in Liberia, Guinea, and Sierra Leone, Ebola traveled to Nigeria and Mali, to Spain, Germany, and the United States. It struck doctors and nurses in unprecedented numbers, wiping out a global health system that was weak in the first place. One August day in Liberia, six pregnant women lost their babies when hospitals couldn't admit them for complications, multiple fatalities in a single day. Anyone willing to treat Ebola victims ran the risk of becoming one.

4.16.3   Detectability—prevent escape

It seemed that little could be done to stop the disease from spreading further. Governments were ready to mitigate the risk, as was the World Health Organization (WHO), which was in denial and snarled in red tape. The world's response to date hasn't fully utilized the statistical risk engineering big data tools that could play a vital role in both protecting health workers from exposure and stemming broader spread of the epidemic virus in the United States, Europe, and elsewhere.

4.16.4   How to reduce/mitigate Ebola risk?

Like Dr. John X. Wang described in his book *What Every Engineer Should Know About Risk Engineering and Management,* for the Art of War against failures, we need to develop protocols based on the Defense in Depth philosophy. To protect the health workers on the front lines in the fight against Ebola, statistical methods and statistical risk engineering big data methods will enable us to analyze proactively rather than reactively. Statistical risk engineering big data should be a core component of our strategy to help us to answer the following critical questions:

- How can health workers be better protected?
- How can the spread of Ebola be slowed globally?
- How can we eventually stop Ebola in the general population?

Like *TIME* commented, "Ebola is a war, and a warning. The global health system is nowhere close to strong enough to keep us safe from infectious disease, and 'us' means everyone."

Yes, "us" means everyone including every engineer.

4.17   Women are from Venus? Men are from Mars? Universal risk engineering?

According to Dr. John Gray, "Men are from Mars, women are from Venus." While everything is possible, from the risk engineering perspective, this is definitely a challenging journey for both ladies and gentlemen, considering the temperatures for Venus and Mars. Included in Figure 4.13 are (from top to bottom) images of Mercury, Venus, Earth (and moon), Mars, Jupiter, Saturn, Uranus, and Neptune.

4.17.1   Venus: Risk engineering based on Arrhenius model?

While Venus is very nearly a twin of Earth in mass and radius, radio and infrared observations show that Venus' surface is heated to 750 Kelvin (891°F), much hotter than Earth.

Figure 4.13 Included are (from top to bottom) images of Mercury, Venus, Earth (and moon), Mars, Jupiter, Saturn, Uranus, and Neptune. (Courtesy of National Aeronautics and Space Administration.)

The high temperature is the result of a deep atmosphere coupled with an immense greenhouse effect. Creating environmentally sensible products is thus critical to mitigate the risk of global warming over Venus.

Moreover, Venus remains the same temperature at night. We can apply the Arrhenius model to predict take-off reliability from Venus' surface.

4.17.2 Mars: Low-cycle fatigue reliability modeling?

While its day is almost the same length as Earth, reddish Mars is very cold. The planet temperature reaches at most to the freezing point of water.

Although the permanent ice caps are in part water ice, the temporary caps expand in winter, forming frozen carbon dioxide "snow." Low-cycle reliability modeling seems necessary to predict take-off reliability from the Martian surface.

The spare clouds are made of both water and dry ice, a great challenge to the mission reliability of our flight.

4.18   Ecosystem risk engineering

Ecosystem risk engineering ensures that Earth's ecosystems remain sustainable such that we continue to enjoy the common resources like air, soil, water, and wildlife.

Ecosystem decision making under uncertainty has emerged from the pragmatic need to value the natural environment and its component species. Some species, such as the polar bear of the Arctic, may be seriously endangered because of its dependency on the isolated shelf ice on which to hunt seals and give birth (see Figure 4.14).

Research shows that natural ecosystems provide the profound societal economic gain than do ecosystems converted from narrow immediate objectives.

Ecosystems are one of the largest untapped resources at our disposal. By designing and embedding smart ecosystems, we can harvest valuable functional uses, increase the performance of buildings and cities, and design more inspiring environments.

An ecosystem industrial design engineer develops custom ecosystems for use in urban landscapes, agriculture, buildings, and industry.

Figure 4.14 Melting Arctic Circle ice drives polar bears closer to extinction. (Courtesy of National Aeronautics and Space Administration, http://climatekids .nasa.gov/review/arctic-animals/polar_bear_clinging-250.jpg.)

4.18.1   Can you "design" an ecosystem?

Yes, ecosystems can be designed to perform a wide variety of functions by combining the necessary ingredients and materials, and tuning them to their environment. When developed and integrated well, ecosystems can have large beneficial effects on all environments. Examples of ecosystems are:

- Edible rooftop gardens
- Urban water filtration systems
- Ecologically activated parklands and urban landscapes
- Urban redevelopment systems
- Ecological sound barriers and indoor climate conditioning
- Social gathering and education functions
- Bio-fuel harvest parks
- High performance organic agriculture (Polydome)

4.18.2   Why use ecosystems for functions?

There are various reasons why you would want to use an ecosystem to perform functional tasks. The most important reason is that ecosystems always have a large set of positive side effects. Unlike a machine that performs a single function, ecosystems can deliver all their benefits at the same time, and the more they are tapped into, the better they can work. This way they can deliver more value on the whole than mechanical systems.

Also, they look amazing, people are strongly attracted to them, and they generally tend to increase the quality of living environments. When designed well, ecosystems last longer, are more resilient, and perform better than any other solution.

4.18.3   How do you design an ecosystem?

We can design ecosystems by bringing together biologists, ecologists, engineers, and designers. Together they research the specific qualities that are needed for the local climate and functional conditions.

Based on this, we can make a smart combination of plants, microorganisms, materials, and design features to maximize the required qualities. The outputs can range among the following qualities, and beyond:

- Food production
- Water filtration and retention
- (Air) particulate filters
- Social functions, gathering, and tourism

- Education
- Energy generation
- Material fabrication (medical, construction, etc.)
- Elevated concentration and health benefits
- Biodiversity increases

4.19 Environmental art and ecological risk engineering: Grand River Sculpture

Environmental art covers a broad range of artistic practices encompassing both historical approaches to nature in art and more recent ecological risk engineering types of works.

For example, in Grand Rapids, Michigan, Grand River Sculpture and Fish Ladder was created to assist the fish in their struggle going upstream, and also provide a visual component of this situation to the public (see Figure 4.15).

Set above the concrete and steel ladder where spawning fish are aided in their upstream battle, an angular structure provides viewing spaces for those who come to observe the Grand River and surrounding environment.

Grand River at Grand Rapids near the Fish Ladder downtown. Stage = 14.7'

Figure 4.15 Grand River at Grand Rapids—Looking downstream from Gage (11) Sixth Street Dam near the Fish Ladder. (Courtesy of National Weather Service.)

Having been "dedicated on June 6, 1975 as a statement of the aesthetic scientific significance of our environment," this piece of environmental art functions both as a usable ladder for the aquatic travelers of the river and an attraction for residents and visitors.

Companies that think in terms of a "product ecosystem" will trounce rivals that come to the market with a more myopic view of success. To create product ecosystems, companies may have to leave behind their comfortable ways of thinking, which may be challenging.

For example, Polydome is a revolutionary approach to greenhouse agriculture that offers the possibility of commercial scale, net-zero-impact food production. The Polydome system strategically interweaves a wide variety of crops and animals, taking advantage of every inch of the greenhouse while eliminating the need for synthetic fertilizers and pesticides. With its high yields (60–90 kg per square meter) and diverse outputs (over 50 crops, two mushroom varieties, chickens, eggs, fish, and honey), even a small Polydome system can provide a richly varied food supply for a large population. It is estimated that by using Polydome, even cities as densely populated as New York City could provide the majority of their own food supply using available roof space. A less dense city like Rotterdam could provide an estimated 80% of its food needs using only 3% of its surface area.

4.20   *Plunging like a polar bear: Mitigating risk and uncertainty in Alaska's ice waters*

Alaska, "The Last Frontier," where I received a "Certificate of Achievement" during a cruise in 2015 from Holland America Line "for braving the waters and participating in the Polar Bear Plunge" (see Figure 4.16).

Figure 4.16 Plunging like a polar bear.

How to swim and plunge like a polar bear? Here is what we can learn from the polar bears for mitigating risk and uncertainty in Alaska's ice waters:

1. Polar bears are good swimmers and have been reported swimming at least 40 miles.
2. Polar bears can dive under ice and use seal haul-out holes to surface in.
3. Polar bears swim rather high in the water, with the head and shoulders out of the water.
4. They swim using the fore-feet only, with the hind legs trailing and the long neck stretched out.
5. Polar bears use their front legs for propulsion in the water, with the hind legs acting as a rudder.
6. Polar bears can remain underwater for 2 minutes.
7. Polar bears are good swimmers and divers. They swim with the hind legs trailing, using the forelegs only for propulsion. The muzzle is extended, but in rough seas the head (including muzzle and eyes) may be submerged and the muzzle rise only for the bear to breathe.
8. Polar bears are strong swimmers; all age and sex classes swim voluntarily, but younger individuals may spend more time in the water, and engage in swimming without any apparent purpose.
9. Cubs sometimes sprint across the ice and dive head-first into the water, similar to the actions of older bears rushing at a seal.
10. To dive, they usually go down head-first, with the hind end initially out of water.
11. When stalking, they use a different method, surfacing with the nose only, and then slipping back under the water.
12. Polar bears have excellent control while swimming, such that they can surface, breathe, raise their head to look around, and submerge again without causing the water to ripple.
13. They swim under the ice between breathing holes during some aquatic hunts, while in other hunts swim along interconnecting channels through the ice.
14. Bears have been observed to spend several hours swimming in one direction (about 5 miles travelled in this way in one instance).
15. Submergence times of 37–72 seconds were recorded.

In summary, polar bears are good swimmers, using their front legs for propulsion and their back legs as a rudder. They can swim distances as long as 65 km over open water. They can dive and swim under water for up to 2 minutes. Usually they dive head-first; however, when stalking in the water they just raise their head or nose out of the water then slip back under the surface. Swimming and plunging like a polar bear, we can mitigate the risk and uncertainty of diving into the ice water. Remember: Safety Always Comes First!

chapter five

Cellular manufacturing
Mitigating risk and uncertainty

Engineering is the art or science of making practical.

Samuel C(harles) Florman
The Existential Pleasures of Engineering (1976)

5.1 Cellular manufacturing of "the invisible" at nanoscale

Over the past few decades, advances in science and technology have enabled manufacturers to smash the limitations they previously faced. Historically, cellular manufacturing has been limited to the things we could physically see and touch, but thanks to nanotechnology, this is no longer the case.

Now, consumer goods are lighter than ever, sportswear and equipment are more aerodynamic and flexible and advanced packaging enables food and drinks to stay fresher for longer. All of this is possible through nanotechnology.

Nanotechnology describes the branch of technology that deals with modeling, measuring, manipulating, or creating matter on a minute scale—between 1 and 100 nanometers, in fact. For manufacturers, nanotechnology is predicted to be a main driver for business in this century, impacting almost every manufacturing sector, from food packaging to consumer goods. But the potential of nanotechnology moves far beyond keeping your bottled cocktail and beer fizzy.

Cellular manufacturing is implemented for "the invisible" at nanoscale. Talking about food packaging, some manufacturers have introduced nanoscale polymers designed to prevent oxygen from leaking through food packaging and spoiling the product inside. This technology is now being widely used to prolong the shelf life of fresh food.

In the United States, a plastic beer bottle was recently introduced that uses nanoparticles of clay to fill up space in the walls of the bottle. These nanoparticles make it harder for carbon dioxide to escape from the beer, keeping the fizz in the beverage. While these innovations are impressive, nanotechnology goes far beyond fixing lifeless lager.

Nanomanufacturing describes manufacturing at the nanoscale of between 1 and 100 nanometers. To put that into perspective, the thickness of a sheet of paper is around 100,000 nanometers. This tiny scale has enabled what is known as atomically precise manufacturing (APM). APM has already opened doors to innovation in design and manufacturing. While APM unlocks a world of potential for manufacturers, what does this mean for the manufacturing process?

APM, ultra-precise automation, and robotics have already proven capable of achieving incredibly accurate results for manufacturers, but there are possibilities even beyond this. The fact that food packaging and material goods have been fabricated atom-by-atom leads to speculation that tiny nanoscale machines could be a possibility in the near future. These nanoscale machines could be used to manufacture materials on an atom-by-atom scale, therefore facilitating an endless amount of possibilities.

But for now, this is all speculation. There are a number of factors that will influence whether nanotechnologies will be integrated into standard industry process. In fact, economic, social, or technical issues will all have an impact.

APM is leading the way and is impacting everything from consumer goods to medicine, enabling cellular manufacturing of "the invisible" at nanoscale, mitigating risk and uncertainty atomically precisely.

5.2   Cellular manufacturing: Mitigating risk and uncertainty with nano solar assembly

The unrestrained combustion of fossil fuels has resulted in vast pollution at the local scale throughout the world, while contributing to global warming at a rate that seriously threatens the stability of many of the world's ecosystems. There are many applications where heat, not electricity, might be the desired outcome of solar power. For example, in large parts of the world the primary cooking fuel is wood or dung—which produces unhealthy indoor air pollution, and can contribute to deforestation. Solar cooking could alleviate that—and since people often cook while the sun isn't out, being able to store heat for later use could be a big benefit. Unlike fuels that are burned, this system uses material that can be continually reused. It produces no emissions and nothing gets consumed.

As shown in Figure 5.1, new solar panel films incorporate nanoparticles to create lightweight, flexible solar cells. Solar photovoltaic (PV) technology is a clean, sustainable, and renewable energy conversion technology that can help meet the energy demands of the world's growing population. Although PV technology is mature with commercial modules obtaining over 20% conversion efficiency, there remains considerable

Figure 5.1 New solar panel films incorporate nanoparticles to create lightweight, flexible solar cells. (Courtesy of Nanosys and NIST.)

opportunities to improve performance. The nearly global access to the solar resource coupled with nanotechnology innovation-driven decreases in the costs of PV provides a path for a renewable energy source to significantly reduce the adverse anthropogenic impacts of energy use by replacing fossil fuels. Current research explores several approaches to improving PV efficiency with nanotechnology:

- Optical enhancement.
- Micro-structural optimization for electronic material quality and increasing the spectral response via band-gap engineering.
- Nano-antennae being tuned into infrared energy, which is radiated down on us all day by the sun and re-radiated back from the Earth at night. These solar collectors don't need their beauty sleep like regular solar cells, with potential efficiencies of up to 80%! They pull overtime, day in and day out if researchers can figure out how to convert the energy into something useful by humans.
- Carbon nanotube, which absorbs the sun's heat and stores that energy in chemical form, ready to be released again on demand. For applications where heat is the desired output—whether for heating buildings, cooking, or powering heat-based industrial processes—this could provide an opportunity for the expansion of solar power into a new realm.

Large-scale utilization of solar-energy resources will require considerable advances in energy-storage technologies to meet ever-increasing global energy demands. Other than liquid fuels, existing energy-storage materials do not provide the requisite combination of high energy density, high stability, easy handling, transportability, and low cost. New hybrid solar thermal fuels, composed of photoswitchable molecules on rigid, low-mass nanostructures, transcend the physical limitations of molecular solar thermal fuels by introducing local sterically constrained environments in which interactions between chromophores can be tuned. A hybrid solar thermal fuel can be produced using azobenzene-functionalized carbon nanotubes (CNTs). On composite bundling, the amount of energy can be stored per azobenzene more than doubles from 58 to 120 kJ mol^{-1}, and the material also maintains robust cyclability and stability. Solar thermal fuels composed of molecule–nanostructure hybrids can exhibit significantly enhanced energy-storage capabilities through the generation of template-enforced steric strain.

The working cycle of a solar thermal fuel is depicted in this illustration, using azobenzene as an example. When such a photoswitchable molecule absorbs a photon of light, it undergoes a structural rearrangement, capturing a portion of the photon's energy as the energy difference between the two structural states. When the molecule is triggered to switch back to the lower-energy form, it releases that energy difference as heat. The principle is simple: some molecules, known as photoswitches, can assume either of two different shapes, as if they had a hinge in the middle. Exposing them to sunlight causes them to absorb energy and jump from one configuration to the other, which is then stable for long periods of time.

Nevertheless, these photoswitches can be triggered to return to the other configuration by applying a small jolt of heat, light, or electricity—and when they relax, they give off heat. In effect, they behave as rechargeable thermal batteries: taking in energy from the sun, storing it indefinitely, and then releasing it on demand. In order to reach the desired energy density—the amount of energy that can be stored in a given weight or volume of material—it is necessary to pack the molecules very close together, which proved to be more difficult than anticipated.

The interactions between azobenzene molecules on neighboring CNTs make the material work. While previous modeling showed that the packing of azobenzenes on the same CNT would provide only a 30% increase in energy storage, the experiments observed a 200% increase.

Providing new motivation for researchers to design more and better photochromic compounds and composite materials that optimize the storage of solar energy in chemical bonds, further exploration of materials and manufacturing methods will be needed to create a practical system for production, incorporating cellular manufacturing: mitigating risk and uncertainty.

Similarly, nano-antennae use common ingredients and can be printed on flexible plastics, like bag, covering a sheet of plastic with millions of nanoscale collectors. The highly integrated process requires Cellular Manufacturing to Mitigate Risk and Uncertainty with nano-antennae assembly/installation.

5.3   *Cellular manufacturing of* The Scorch Trials

Maze Runner: The Scorch Trials is a 2015 American dystopian science fiction action thriller film based on James Dashner's novel *The Scorch Trials*, the second novel in *The Maze Runner* book series. After having escaped the Maze, the Gladers now face a new set of challenges on the open roads of a desolate landscape filled with unimaginable obstacles.

Transported to a remote fortified outpost, Thomas and his fellow teenage Gladers find themselves in trouble after uncovering a diabolical plot from the mysterious and powerful organization WCKD. In the movie, as Mary halts Brenda's infection using an enzyme cure, Mary explains that the enzyme can only be harvested from an Immune's body, not manufactured, and that arguments over the methods of manufacturing the cure with Ava lead to Mary's departure from WCKD. Enzymes are biological molecules (proteins) that act as catalysts and help complex reactions occur everywhere in life.

In biology, the network of interacting proteins, commonly called the interactome, really consists of enzymes and protein inputs/outputs to the metabolism. If we could remove from the protein-protein interactome the inputs and outputs leaving a connected graph or a time sequence list of the enzymes that participate in cell cycle and cellular manufacturing, then we would essentially have a linear network of cellular manufacturing. If we constructed what is called the edge graph for the linear network, we would have a network of enzymes, or processing machines, as they are used in sequence.

Systems biology is a domain that generally encompasses both large-scale, organismal systems, and smaller-scale, cellular systems. The majority of contemporary systems biology falls under the cellular-scale studies with the large goals of understanding genome to phenome mapping. This cellular-scale, or molecular systems biology, may also contribute to synthetic biology by becoming the theoretical underpinning of that, largely, engineering discipline; and it may also contribute to a perennial question of physics—the difference between living and non-living matter. There is significant other research focusing on defining the difference between living and nonliving matter. These include:

- Category theory
- Genetic networks
- Complexity theory and self-organization, and autopoiesis
- Turing machines and information theory

While it would take a full-length book to review the many subjects that already come into play in discussing the boundaries between the living and nonliving, here we concern ourselves mainly with factory system analogies and cellular molecular networks, as we explore the boundaries that define life.

At a fundamental level cells, like factory production systems, contain anticipatory systems and much of the mathematics associated with factories can be exploited for systems biology. It is a very common network motif in molecular system networks. An abstract example of the arabinose system of *Escherichia coli* is shown in Figure 5.3. Another example is the MAP kinase cascade. These are known as anticipatory systems and contain within themselves models of the system and the system controller. The phrase anticipatory system, by itself, seems to ignore causality. But in fact the causality is preserved by the fact that the model uses information from prior system states to predict future states. These anticipatory systems are said to be able to anticipate the future, but as we will see, these systems contain implicit system models of process controllers that enable them to seemingly anticipate the future. Because there is no explicit model, the actual process being controlled can drift in performance due to subsystem changes. The only assumptions in this model are that each chemical species is "processed" by a unique enzyme to produce another chemical species.

For cellular systems biology, we can view the system as a network of interacting molecular species, with one of the major time lags being diffusion and Brownian motion. Processes can take place reasonably rapidly and Le Chatelier's principle can drive the system dynamics. On the organism level, diffusion and other transport processes can be major time delays, and the dynamics of the organism can be minutes to days to weeks. Similarly, the time lag in manufacturing is far greater between sensing a manufacturing processing component failure (mean time to failure) and actual repair time (mean time to repair). This gives rise to a hysteresis.

In biology, the network of interacting proteins, commonly called the interactome, really consists of enzymes and protein inputs/outputs to the metabolism. If we could remove from the protein-protein interactome the inputs and outputs leaving a connected graph or a time sequence list of the enzymes that participate in cell cycle and/or cellular manufacturing, then we would essentially have the following type of linear network as shown by Equation 5.1:

$$W(t_1) \xrightarrow{P_{1,2}} W(t_2) \xrightarrow{P_{2,3}} \ldots \xrightarrow{P_{n-2,n-1}} W(t_1) \xrightarrow{P_{n-1,n}} W(t_n) \qquad (5.1)$$

where

- $W(t_i)$ represents the metabolites or materials to be processed by $P_{i,j}$ during the time period between i and j.
- Biochemically $P_{i,j}$ would be enzymes.

In a manufacturing environment, it would be the processing machine.

As shown by Equation 5.2, now if we constructed what is called the edge graph for the linear network shown above, we would have:

$$P_{1,2} \rightarrow P_{2,3} \rightarrow \cdots \rightarrow P_{n-2,n-1} \rightarrow P_{n-1,n} \tag{5.2}$$

a network of enzymes, or processing machines, as they are used in sequence.

We have been drawing several parallels between manufacturing and systems biology. Since manufacturing networks are completely known we have an opportunity to explore algebraic graph theory and test algebraic and group theory hypothesis on manufacturing networks that are not possible yet with incomplete biological interactomes.

In summary, the development of main equations for anticipatory systems and metabolism-repair systems are similar for manufacturing systems and cellular biology. The fact that these two disparate domains are so tightly coupled by similar mathematics suggests that these concepts are indeed at the boundary of the living and nonliving.

Here, we have presented the parallel between cellular manufacturing and systems biology. Since cellular manufacturing networks are completely known we have an opportunity to explore algebraic graph theory and test algebraic and group theory hypothesis on cellular manufacturing networks that are not possible yet with incomplete biological interactomes.

With this new ally—cellular manufacturing—we can help the Gladers to stage a daring escape into the Scorch, a desolate landscape filled with dangerous obstacles and crawling with the virus-infected Cranks, since "THE MAZE WAS JUST THE BEGINNING."

5.4   *The world is flat: Cellular manufacturing in a world which is flat yet spherical*

The author was mentioned by 3D PRINT Injection: "If you're doing any international #manufacturing and #engineering, you'll want to know @JohnJxwang with https://drjohnxwang.wordpress.com."

Actually, if you're doing any international #manufacturing and #engineering, you may also want to know cellular manufacturing and 3D printing, because "The World is Flat":

- Cellular manufacturing arranges factory floor labor into semi-autonomous and multi-skilled teams, or work cells, which manufacture complete products or complex components. Properly trained and implemented cells are more flexible and responsive than the traditional mass-production line, and can manage processes, defects, scheduling, equipment maintenance, and other manufacturing issues more efficiently.

- 3D printing is quickly becoming a very affordable option for producing physical objects. The term "3D printing" covers a number of closely related technologies, all of which produce a 3-dimensional physical object from a computer model by building it up in successive layers. The principle of 3D printing includes laminating manufactured add-in, the printing material in the form of the XY plane depicted by scanning cross-sectional shapes of the 3D model, and then making the Z-axis direction displacement discontinuity single slice thickness, forming 3-dimensional physical.

The typical approach in manufacturing has been subtractive—to remove materials to make the final product useful, for example, to cut or drill materials away, leaving only the desired substance. The traditional manufacturing is subtractive manufacturing areas, namely through the mold, milling, and other machining techniques and tools to convert raw materials into products and processes, characterized by the use of reduced cutting, reducing materials to produce parts.

As shown by Figure 5.2, 3D printing is being widely applied, including 3D printer during testing in the Microgravity Science Glovebox. Compared with the traditional manufacturing, 3D printing does not require complex technology, does not require a huge machine, does not require a large number of manpower, there is no lengthy pipeline operations; the 3D printing is directly through the computer 3D model data, any shape of parts can be produced by the printer, so that a wider range of

Figure 5.2 The 3D printer during testing in the Microgravity Science Glovebox (MSG) Engineering Unit at Marshall Space Flight Center. (Courtesy of NASA.)

products can be manufactured. The desired components can be made of materials that can be sprayed as liquids (e.g., plastics or resins) or formed from melted powder (including metals and ceramics). The components are typically manufactured directly from a 3D CAD image.

The world is flat. 3D printing enables us to model elliptic geometry, the geometry of navigating our stars. The simplest model for elliptic geometry is the surface of the earth with the lines of longitude on its surface. If you and a friend both move north you would be on parallel lines if the world is flat but since it is a ball you two will meet at the North Pole. For the challenge of globalization, cellular manufacturing is increasingly controlled by computer-aided design (CAD) tools, creating 3D printing— visualize our world which is flat yet spherical.

5.5 Bottled cocktail: Cellular manufacturing in the beverage industry

Bottled cocktails are quickly being adopted by a number of innovative restaurants/bars, like the West Restaurant in Vancouver, allowing their bartenders to establish a rapport with their customers, while serving a zero-defect drink.

Indeed, the beverage industry is in full swing with innovation. A variety of bars are favoring days spent filling bottle after bottle of pre-batched cocktails, much like a commercial production line or bottling plant.

Batching cocktails allows the bartender the chance to completely control the repeatability and reproducibility of their serve, which is balanced each and every time. It also allows the bartender to finely tune the cocktail. Dashes of potent ingredients, such as absinthe, can be carefully measured in easier quantities. It reduces cycle-time, allowing the bartender more time with his or her customers.

What are the benefits to the business bottom-line of a pre-batched cocktail bar? Here are two major advantages:

- First, removing any perishables leads to a business with near-to-no waste, whereas a standard cocktail bar could be factoring in around 0.5% to 1% of sales in waste alone.
- The customer also benefits from consistency, a key feature of any successful bar. The ratio of spirits, acids, and salts are perfectly balanced every time and both diluted and chilled to perfection. The guests are guaranteed that their drink will taste exactly as it had on their previous visit, something that is very difficult to provide in a conventional bar environment.

In summary, the practice of pre-batching cocktails could provide us with the ultimate bar; almost robotic and precise to the point of being flawless. It brings an increased interaction between bartender and guest, balanced cocktails and ludicrous speed of service.

5.6   Integrate TRIZ into cellular manufacturing to improve productivity

With increasing competition in the market, expediting the problem solving process has become crucial in the industry. The reduction of waste is essential in the present scenario and to survive in the competitive market, the appropriate concept has been taken as cellular manufacturing. As such, a number of problem solving techniques have been devised to efficiently tackle problems of varying natures. The theory of inventive problem solving (TRIZ) has been widely applied in a variety of industries and services recently.

Integrating TRIZ into cellular manufacturing, for example, a manufacturing system composed of several cells, as shown in Figure 5.3, improves productivity. For cellular manufacturing, a synergistic approach will be appreciable for which TRIZ has been utilized to develop the solution to improve the productivity. Cellular manufacturing involves rearranging traditional operation-based factory layouts into process-based cells that promote a smooth production flow by culling waste. Cellular manufacturing has been applied to find and narrow down the problems while TRIZ can be applied to solve the contradictions and achieve better results.

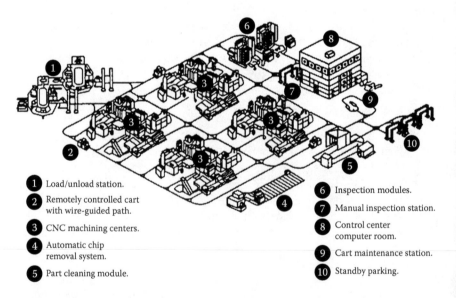

1 Load/unload station.

2 Remotely controlled cart with wire-guided path.

3 CNC machining centers.

4 Automatic chip removal system.

5 Part cleaning module.

6 Inspection modules.

7 Manual inspection station.

8 Control center computer room.

9 Cart maintenance station.

10 Standby parking.

Figure 5.3 Manufacturing system composed of several cells.

In general, the result obtained by the application of the synergistic concepts will give more benefits than that of the individual concepts. The book titled *Cellular Manufacturing: Mitigating Risk and Uncertainty* introduces engineering teams to basic cellular manufacturing and teamwork concepts, and orients them for participating in the design of a new production cell.

5.7 Optimize topology for networked virtual cellular manufacturing (NVCM)

Virtual cellular manufacturing (VCM) "combines the setup efficiency typically obtained by group technology (GT) cellular manufacturing (CM) systems with the routing flexibility of a job shop."

For VCM, network topology is the arrangement of the various elements (links, nodes, etc.) of a production network.

As shown by Figure 5.4, a networked virtual cellular manufacturing (NVCM) system incorporates network topology into traditional VCM systems, measuring and reporting in real time both local and geographically remote distributed VCM processes.

As an artificial intelligence (AI) technique that optimizes a problem by iteratively trying to improve a candidate solution with regard to a given measure of utility, particle swarm optimization (PSO) is a population-based stochastic search algorithm.

Inspired from observations of birds flocking and fish schooling, PSO optimizes a problem by having a population of candidate solutions, here dubbed particles, and moving these particles around in the search-space

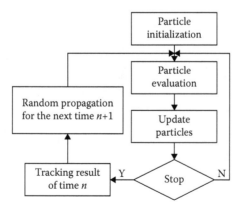

Figure 5.4 Optimize topology for networked virtual cellular manufacturing (NVCM).

according to simple mathematical formulae over the particle's position and velocity with the following simple optimization procedure:

```
Initialize parameters
Initialize populations
Evaluate
Do {
Find the personal best
Find the global best
Update velocity
Update position
Evaluate
}While(End)
```

Comparing with individual virtual cellular manufacturing systems, a networked virtual cellular manufacturing system is population-based. PSO can be applied to optimize the population topology of a networked virtual cellular manufacturing system.

5.8   Euchre and cellular manufacturing: Mitigating risk and uncertainty

"Play is the highest form of research," Dr. Albert Einstein said. Play is also the highest form of engineering. As a popular card-game in the Midwest, especially in Michigan and Ohio, Euchre manifests how we mitigate risk and uncertainty, a central theme of cellular manufacturing.

Euchre is a risk-taking card game most commonly played with four people in two partnerships with a deck of 24, or sometimes 32, standard playing cards. It is the game responsible for introducing the joker into modern packs. Originated in the Alsace region of France near the border of Germany from a game called Juckerspiel, Euchre was invented around 1860 to act as a top trump or best bower (from the German word Bauer, "farmer," denoting also the jack). In German, Jucker means "jack" and spiel means "game." It is believed to be closely related to the French game Écarté that was popularized in the United States by the Cornish and Pennsylvania Dutch, and to the seventeenth-century game of bad repute Loo. It may be sometimes referred to as Knock Euchre to distinguish it from Bid Euchre. Euchre is all about the jacks. In this game, the jack in the trump suit is the most powerful card.

Section 2.3 "Decision under Uncertainty" of the book titled *Cellular Manufacturing: Mitigating Risk and Uncertainty* helps to develop an optimal policy, based on Markov (or semi-Markov or non-Markov) decision theory for the capacity management problem in a manufacturing firm facing stochastic market demand. The manufacturing firm could implement a reconfigurable manufacturing system and face a delay between the times capacity changes are ordered and the times they are delivered. Here, optimal policies are presented as optimal boundaries representing the optimal capacity expansion and reduction levels. To increase the robustness of the optimal policy to unexpected events, as explored by Chapter 6 "Robust Design of Cellular Manufacturing Systems" of the book titled *Cellular Manufacturing: Mitigating Risk and Uncertainty*, the concept of feedback control is applied to address the capacity management problem. It is shown that feedback provides suboptimal solutions to the capacity management problem which are more robust under unexpected disturbances in market demand and unexpected events.

The cellular manufacturing community is faced with an extensive amount of data. Software programs are being developed to examine this issue of data overload and to develop solutions. The responsibility of making the final software decision lies on the analyst. Therefore, the interface is the key to linking the intelligence data to the processing and results. If the interface is difficult and complex, the software will be less likely to be used. A methodology must be created which can objectively evaluate the effectiveness of the interface. This methodology will also measure the improvements in the interface's effectiveness that results when various changes are made to the original software interface. Value stream mapping (VSM) is a proven methodology that can be applied to this problem. VSM provides an objective methodology to identify the values of an organization. Its hierarchical structure is well suited for handling multi-objective problems, such as identifying the values of software interfaces. The values can be measured and put to a common scale, allowing their contribution to the overall objective to be evaluated. By assigning quantifiable measurements to the components, the multi-objective goal can be evaluated and insight can be provided to the decision makers involved with the intelligence software. VSM was applied to determine what is valued in software's interface to members of the cellular manufacturing community. With these values identified, software that is under development was evaluated against the hierarchy. This provided insight into where improvements could be made to the interface that would provide the greatest benefit. The VSM process also allows for the decision maker to continually reevaluate the software against the hierarchy, enabling continual improvement on the interface while maintaining the values of the cellular manufacturing community.

Play is the highest form of engineering. When you visit "the beautiful Grand River by which my sweet home is located," you consider playing Euchre, a popular card-game by the great river.

5.9 Psychology regression: Why is the past entangled with the present?

Entanglement is a term used in quantum mechanics to describe the way that particles of energy/matter can become correlated to interact with each other regardless of how far apart they are.

As revealed by *The Gift*, an Australian-American psychological thriller, the past could entangle with the present due to psychology regression. In the movie, a young married couple, Simon and Robyn, are living what appears to be a picture-perfect life. However, their lives are thrown into a harrowing tailspin when Gordo, an acquaintance from Simon's past, brings mysterious gifts and a horrifying secret to light after more than 20 years.

From psychology's perspective, why is the past entangled with the present? According to psychological risk engineering, *The Gift* tackles the bigger idea about the past's role in the present. Where is the root cause of Simon's insecurities? Why would he always worry that his endlessly supportive wife, Robyn, may be too good for him? Would the insecurities and worry be related to his past, the horrifying secret of more than 20 years ago?

It is similar to the Stockholm syndrome but in this case an individual's past and present environment take the place of an abductor. If we reflect upon the movie *The Boy* discussed in Section 5.11, we can see the doll is just a metaphor there. When men and women are placed in a situation where they no longer have any control over their fate, feel intense fear of physical harm, and believe all control is in the hands of their tormentor, a strategy for survival can result which can develop into a psychological response that can include sympathy and support for their captor's plight.

The name Stockholm syndrome was derived from a 1973 bank robbery in Stockholm, Sweden, where four hostages were held for six days. Throughout their imprisonment and while in harm's way, each hostage seemed to defend the actions of the robbers and even appeared to rebuke efforts by the government to rescue them. Months after their ordeal had ended, the hostages continued to exhibit loyalty to their captors to the point of refusing to testify against them, as well as helping the criminals raise funds for legal representation.

The response of the hostages intrigued behaviorists. Research was conducted to see if the Kreditbanken incident was unique or if other hostages in similar circumstances experienced the same sympathetic, supportive bonding with their captors. The researchers determined that such behavior was very common. Other famous cases included:

- On June 10, 1991, witnesses said they saw a man and a woman abduct 11-year-old Jaycee Lee Dugard by a school bus stop near her home in South Lake Tahoe, California. Her disappearance remained unsolved until August 27, 2009, when she walked into a California police station and introduced herself.

For 18 years she was held captive in a tent behind the home of her captors, Phillip and Nancy Garrido. There Ms. Dugard gave birth to two children who were ages 11 and 15 at the time of her reappearance. Although the opportunity to escape was present at different times throughout her captivity, Jaycee Dugard bonded with the captors as a form of survival.

- Another more famous case in the United States is that of heiress Patty Hearst, who at age 19 was kidnapped by the Symbionese Liberation Army (SLA). Two months after her kidnapping, she was seen in photographs participating in an SLA bank robbery in San Francisco. Later a tape recording was released with Hearst (SLA pseudonym Tania) voicing her support and commitment to the SLA cause.

After the SLA group, including Hearst, was arrested, she denounced the radical group. During her trial her defense lawyer attributed her behavior while with the SLA to a subconscious effort to survive, comparing her reaction to captivity to other victims of Stockholm syndrome. According to testimony, Hearst was bound, blindfolded, and kept in a small dark closet where she was physically and sexually abused for weeks prior to the bank robbery.

What causes Stockholm syndrome? Individuals can apparently succumb to Stockholm syndrome under the following circumstances:

- Believing one's captor can and will kill them.
- Isolation from anyone but the captors.
- Belief that escape is impossible.
- Inflating the captor's acts of kindness into genuine care for each other's welfare.

Victims of Stockholm syndrome generally suffer from severe isolation and emotional and physical abuse demonstrated in characteristics of battered spouses, incest victims, abused children, prisoners of war, cult victims, and kidnapped or hostage victims. Each of these circumstances can result in victims responding in a compliant and supportive way as a tactic for survival.

Linked with decision making under uncertainty, regression is a defense mechanism that causes a person to adopt certain behavior traits from an earlier stage of development. Beyond traditional applications in cellular manufacturing and engineering decision making, regression analysis can be applied to mitigate risk and uncertainty in a group setting of daily life.

5.10   *How to regress on Halloween night?*

How to regress on Halloween night? In modern flourishing cities, crowds, industries, and techniques, would people become disconnected from the past?

Halloween originally emerged in the British Isles, where the late fall was gray and gloomy. Would it start a dead season? Would it begin the season of loss—of birds, flowers, leaves, and the warmth of the sun?

When the world is covered with dark gray, would Earth freeze to death as November comes?

Madison J. Cawein's poem "The Wood Water" (1905) may help us to develop a regression plot tonight, a Halloween night.

> An evil, stealthy water, dark as hate,
> Sunk from the light of day,
> 'Thwart which is hung a ruined water-gate,
> Creeps on its stagnant way.
>
> Moss and the spawny duckweed, dim as air,
> And green as copperas,
> Choke its dull current; and, like hideous hair,
> Tangles of twisted grass.
>
> Above it sinister trees,—as crouched and gaunt
> As huddled Terror,—lean;
> Guarding some secret in that nightmare haunt,
> Some horror they have seen.
>
> Something the sunset points at from afar,
> Spearing the sullen wood
> And hag-gray water with a single bar
> Of flame as red as blood.
>
> Something the stars, conspiring with the moon,
> Shall look on, and remain
> Frozen with fear; staring as in a swoon,
> Striving to flee in vain.
>
> Something the wisp that, wandering in the night,
> Above the ghastly stream,
> Haply shall find; and, filled with frantic fright,
> Light with its ghostly gleam.

Something that lies there, under weed and ooze,
With wide and awful eyes
And matted hair, and limbs the waters bruise,
That strives, yet can not rise.

5.11 The Boy: *The game of mitigating psychological risk and uncertainty*

In a psychological thriller, there has to be major elements of suspense, even terror or horror. There are events that drive the protagonist to question everything about his or her ability to succeed. The key of a psychology thriller story is the psychological exploration of each character.

The Boy is a frightening thrill ride presented by a 2016 American horror movie released on January 22, 2016. In the movie, Greta is a young American woman who takes a job as a nanny in a remote English village, only to discover that the family's 8-year-old is a life-sized doll that the parents care for just like a real boy, as a way to cope with the death of their actual son 20 years prior. After violating a list of strict rules, a series of disturbing and inexplicable events bring Greta's worst nightmare to life, leading her to believe that the doll is actually alive.

From a psychological risk engineering perspective, *The Boy* is a horror movie where the real horror is kind of the idea of coping with loss and ignoring reality. Greta, as well as the parents in the film, the Hilshires, are both burying their heads in the sand to some degree. When she gets to the house, it's a really strange situation; however, we also realize she doesn't have a whole lot of other choices and it's really only by letting herself be completely absorbed that she feels safe. As we understand it in the movie, there are those rules that the doll dictates as to how he should be taken care of, and it's only when she abides by these rules that she feels protected and safe. So that brings up interesting questions such as:

- What sort of game is this playing for mitigating risk and uncertainty?
- How do you define safety for yourself?
- How does Greta create her own psychological safety (perceived safety)?

In a psychological thriller, characters shouldn't win using physical talents or efforts, but rather using their mind, wits, or smarts. Greta uses Brahms' own rules against him when she confronts him, demanding that he go to sleep. After Brahms enters his bed, he pulls Greta in for his "good night kiss" when Greta stabs him. She escapes; rescuing Malcolm and leaving the Hillshire estate gate open behind her. Here, actions and psychological aspects both are used to mitigate risk and uncertainty.

5.12   Shanghai stampede tragedy: Need risk engineering for traffic risk control

On December 31, 2014, a deadly stampede occurred in Shanghai, near Chen Yi Square on the Bund, where around 300,000 people had gathered for the New Year celebration. Thirty-six people were killed and there were 49 injured, 13 seriously. As of January 5, 2015, 10 remained in a serious condition. Traffic risk in urban areas has resulted in an increased focus on public safety, economic, financial, and travel reliability impacts.

While investigations continue into the Shanghai stampede tragedy, eyewitness accounts and media reports point to a sequence of miscalculation, miscommunication, inadequate enforcement, and insufficient emergency response by city officials that helped create the out-of-control conditions leading to the stampede.

- Miscalculation: Authorities underestimated the crowd. On the 2014 New Year's Eve, a 300,000-strong crowd filled the city's famed riverfront for the annual midnight light show, far exceeding government expectations.
- Miscommunication: The city had already canceled the traditional light show on the Bund, as the riverfront area is known. However, authorities failed to notify the public effectively. Nevertheless, the new venue also has a similar name to the Bund, which may have added to the miscommunication.
- Inadequate enforcement: The venue was guarded by only 700 police officers with no traffic control to an elevated viewing platform reachable through staircases and the nearest subway station was closed to rein in the crowd.
- Lack of crash protection: No protective measures to control the speed of crowd flow. No effective feedback about what was happening on the scene.
- Insufficient emergency response: Lack of ambulances, hospitals, doctors, and nurses to rescue.

As shown by Table 5.1, the modern engineering approach to traffic risk control began with Dr. William Haddon, who in the late 1960s developed the first systematic method of identifying a complete range of options for reducing traffic risk impacts.

Table 5.1 Incorporate risk engineering into traffic risk control

Phase	Nature of intervention	Human	Factors		
			Facility/infrastructure	Environment	
Pre-crash	Crash prevention	Information/risk communication	Facility-worthiness	Stair design, signs	
		Crowd prediction (stress) Enforcement resources (strength for crash prevention)	Lighting Damping	Markings, maintenance Emergency facilities	
		Flow management	Emergency facilities		
Crash	Injury prevention	Use of restraints during crash	Crowd restraints	Crash-protective	
		Injury prevention/mitigation	Risk communication equipment (with crowd)	Hazardous objects	
		Other risk engineering devices			
		Crash-protective design (strength for crash protection)			
Post-crash	Life-sustaining	First aid skill	Ease of access	Rescue facilities	
		Access to medical care	Collapse risk control	Stampede prevention, congestion mitigation	
				Ambulances and hospitals, etc. (strength for crash mitigation)	

5.13 The opening of Suez Canal: Gateway to risk engineering

One hundred forty-seven years after its construction, the Suez Canal remains one of the most impressive and important thoroughfares on the planet. The 163-km-long canal linking the Mediterranean and Red Seas is as important as a trading route today as it was when it was first opened in 1869.

Although the Suez Canal wasn't officially completed until 1869, there is a long history of interest in connecting both the Nile River in Egypt and the Mediterranean Sea to the Red Sea. It is believed that the first canal in the area was constructed between the Nile River delta and the Red Sea in the thirteenth century B.C.E. During the 1000 years following its construction, the original canal was neglected and its use finally stopped in the eighth century.

The first modern attempts to build a canal came in the late 1700s when Napoleon Bonaparte conducted an expedition to Egypt. He believed that building a French controlled canal on the Isthmus of Suez would cause trade problems for the British as they would either have to pay dues to France or continue sending goods over land or around the southern part of Africa. Studies for Napoleon's canal plan began in 1799 but a miscalculation in measurement showed the sea levels between the Mediterranean and the Red Seas as being too different for a canal to be feasible and construction immediately stopped.

The next attempt to build a canal in the area occurred in the mid-1800s when a French diplomat and engineer, Ferdinand de Lesseps, convinced the Egyptian viceroy Said Pasha to support the building of a canal. In 1854, Ferdinand de Lesseps, the former French consul to Cairo, secured an agreement with the Ottoman governor of Egypt to build a canal 100 miles across the Isthmus of Suez. An international team of engineers drew up a construction plan, and in 1856 the Suez Canal Company was formed and granted the right to operate the canal for 99 years after completion of the work.

As a gateway to risk engineering, construction of the Suez Canal witnessed tremendous schedule risk and cost risk. Construction of the Suez Canal officially began on April 25, 1859, and at first digging was done by hand with picks and shovels wielded by forced laborers. Later, European workers with dredgers and steam shovels arrived. Labor disputes and a cholera epidemic slowed construction, and the Suez Canal was not completed until 1869—four years behind schedule, at a cost of $100 million.

On November 17, 1869, the Suez Canal was opened to navigation. The Suez Canal, connecting the Mediterranean and the Red Seas, was inaugurated in an elaborate ceremony attended by French Empress Eugénie, wife of Napoleon III.

Almost immediately after its opening, the Suez Canal had a significant impact on world trade as goods were moved around the world in record time. When it opened, the Suez Canal was only 25 feet deep, 72 feet wide at the bottom, and 200 to 300 feet wide at the surface. Consequently, fewer than 500 ships navigated it in its first full year of operation. Major improvements began in 1876, however, and the canal soon grew into one of the world's most heavily traveled shipping lanes.

5.14 How to prevent a ping pong ball from "falling back"

How do we keep a ping pong ball suspended in mid-air, to mitigate the risk of "falling back"? We will need a hand-held hair dryer with a hose that can be attached. The dryer will blow air out instead of sucking it in. Here is how to set up your experiment for risk engineering:

- Turn on the dryer; hold it so that it blows straight up.
- Drop a ping pong ball into the center of the air stream.
- Adjust the power to maintain appropriate air flow.
- The ball will dance around, being suspended in the air.

With a little practice, we will be able to move the test setup around the room without the ping pong ball "falling back."

Similar to model-test of wind-blowing, as shown by Figure 5.5, we can fly a ping pong ball because the air stream goes over a curved surface and thus produces an area of lower pressure. According to Bernoulli's principle, the pressure differential pulls the ping pong ball in the direction of the lower pressure.

Figure 5.5 The ping pong ball remains in the stream of air due to lower pressure created around the surface of the ball. (Courtesy of National Oceanic and Atmospheric Administration.)

This is similar to how we prevent London Bridge from falling down. Aerodynamics is the study of the properties of moving air, which is a main component in the design of automobiles, bridges, heating and ventilation, gas piping, and spacecraft. Hydrodynamics is the study of forces exerted on or by fluids, which is the main component in naval architecture or ship design, and ocean engineering. Ocean and marine engineers are responsible for the study of the offshore environment in order to design oil rigs and production platforms as well as floating vessels and subsea pipeline systems needed in the oil production process. Hydraulic engineers use hydraulics, or the use of liquid power to do work, to design heavy machinery, water distribution systems, sewage networks, storm water management systems, bridges, dams, channels, canals, and levees. Various submersibles and remotely operated vehicles designed by engineers are widely used by government and scientific researchers and are essential in the discovery of deep-water communities and the exploration of the abysmal ocean because they can reach depths much greater than previous satellite and shipboard technologies.

5.15   Dancing with the fire: Entertaining risk engineering

Fire dancing (also known as fire twirling, fire spinning, fire performance, or fire manipulation) is a group of performance arts or disciplines that involve manipulation of objects on fire. Typically these objects have one or more bundles of wicking, which are soaked in fuel and ignited.

Some of these disciplines are related to juggling or baton twirling (both forms of object manipulation), and there is also an affinity between fire dancing and rhythmic gymnastics. Fire dancing is often performed to music. Fire dancing has been a traditional part of cultures from around the world, and modern fire performance often includes visual and stylistic elements from many traditions. Figure 5.6 shows a Native American doing a traditional dance at night by a bonfire.

Warning: Fire dancing is a very dangerous performance art, and fire safety precautions should always be taken.

5.15.1   Firewalking on the sun

For centuries, fire has remained a symbol of life, death, rebirth, and purification in various cultures around the world. It is revered by many as a sacred symbol of life, being a provider of light, warmth, and sustenance. People have danced around it, jumped over it, twirled it around, and expelled it from their mouths...but how exactly does one walk on it? In the following, we'll take a closer look at the fascinating phenomenon

Figure 5.6 Native American doing a traditional dance at night by a bonfire. (Courtesy of National Endowment for the Arts.)

of firewalking to see just how this dangerous and controversial practice actually works, without yielding the walker to a new pair of permanently scarred soles.

Firewalking dates all the way back to 1200 B.C. India, when ancient Hindu sadhus would walk across burning hot coals to exhibit their devotion and strength of spiritual connection. In Zoroastrianism, the world's oldest religion, water and fire represent elements of purification and wisdom, and Zoroastrian rituals usually involve prayers in front of a flame. Born in what would now be modern-day Iran, many Zoroastrian customs have carried over into Iranian culture, reflected in such practices as the Chaharshanbe Suri celebration (literally translated as "Red Wednesday"), which falls on the last Wednesday of the Iranian year and involves jumping over large bonfires to cleanse away sickness and evil while hopefully taking on happiness, enlightenment, and energy.

All across the globe some form of firewalking has permeated nearly every culture. In North America, many Indian tribes such as the Fox, Blackfeet, and Zuni had their own shamanic firewalking traditions also involving purification and healing. In northern Greece, villagers firewalk during a 3-day festival honoring Saint Constantine and Saint Helen, believing that the saints provide them with the power to accomplish the task. Due to the ample availability of it, indigenous Hawaiians known as Kahunas usually practiced lava walking, a tradition prevalent throughout many Polynesian tribes. Other regions of firewalking include Bali, Malaysia, Japan, Sri Lanka, and Pakistan.

Throughout most of these cultures, firewalking is seen as a rite of passage, a test of faith, a cleansing ritual, or even a healing practice. On the great African continent the !Kung bushman of the Kalahari desert walk on fire in their healing ceremonies. They believe walking on fire heats up their internal energy, called n/um, and !Kung priests will walk on it, pick it up, put their heads in it, and rub it all over themselves. It is said that once the body is heated up as hot as the coals, the walker cannot be burned. And the hotter their n/um, the more they will fall into a trance-type consciousness in which all surrounding sickly people can be healed.

The perception of fire as a purifying element is not unwarranted, since fire is required on Earth for all life to thrive. Right here in southern California wildfires are an inherent part of the region, clearing old, dry, dying brush to create room for new growth. In fact, some seed pods of regional plants won't even open and disperse unless they are heated up or ignited. Many of these plants also have leaves coated in flammable oil to help perpetuate fire, with seeds that will germinate with only fire or smoke present.

Today firewalking has evolved into a clichéd workplace teambuilding activity, to garner confidence, trust, and motivation. The modern firewalking movement spread thanks to inspirational speakers such as Tony Robbins, who believe that practices such as firewalking help show people that it's possible to do things that seem impossible. The elimination of fear and the realization of self-empowerment are the principles behind modern firewalking occasions.

A man by the name of Tolly Burkan is credited with introducing firewalking into Western culture, with the founding of the Firewalking Institute for Research and Education in the 1970s. Known as the "father of the global firewalking movement," Burkan developed the first firewalking classes to teach the act to the general public. He spent the 1980s and 1990s working with companies who wished to increase employee productivity and efficiency by creating team-building, confidence-building firewalking retreats, where employees could learn mind-over-matter principles and realize their full potential. Even though firewalking does involve a great deal of courage, the reason for why it is possible lies solely in science and not within any sort of mind-over-matter beliefs.

5.15.2   Fire apparatuses

The various tools used by the fire performance community borrow from a variety of sources. Many have martial sources like swords, staves, poi, and whips, where some seem specifically designed for the fire community. The use of these tools is limited only by the imagination of their users. Some tools lend themselves to rhythmic swinging and twirling, others to martial kata, and others to more subtle use. Some common tools are

- Poi: A pair of roughly arm-length chains with handles attached to one end, and bundle of wicking material on the other.
- Staff: A rod of wood or metal, with wicking material applied to one or both ends. Staves are generally used in pairs or individually, though many performers are now experimenting with three or more staves.
- Fire hoop: Hoop with spokes and wicking material attached.
- Fans: A large metal fan with one or more wicks attached to the edges.
- Fire umbrella: An umbrella that has the cloth removed, with Kevlar tips.
- Fire meteor: A long length of chain or rope with wicks, or small bowls of liquid fuel, attached to both ends.
- Nunchaku: Nunchaku with wicking material, usually at either end.
- Batons
- Fire stick: Like a traditional devil stick, with wicks on both ends of the central stick.
- Torch: A short club or torch, with a wick on one end, and swung like Indian clubs or tossed end-over-end like juggling clubs.
- Fire-knives: Short staves with blades attached to the ends and wicking material applied to the blade. Fire knives are the traditional Polynesian fire implement and have been in use since the 1940s.
- Fire rope dart: A wick, sometimes wrapped around a steel spike, at the end of a rope or chain ranging from 6 to 15 feet long, with a ring or other handle on the opposite end.
- Fire sword: Either a real sword modified for fire, or one specifically built for the purpose of fire shows.
- Chi ball/Fire orb: Two rings or handles with a wick attached between them by a thin wire.
- Finger wands: Short torches attached to individual fingers.
- Palm torches: Small torches with a flat base meant to be held upright in the palm of the hand.
- Fire whips: Lengths of braided aramid fiber tapered to make a bullwhip, usually with a metal handle about 12 inches long.
- Jumblymambas: A triple ended fire object for juggling, twirling, and manipulation.
- Fire cannons: A propane flame effect device; larger ones can shoot a pillar of fire up to 200+ feet in the air, although they usually are mounted to a base or vehicle.
- Fire poofers: Similar to fire cannons, but much smaller and made to be held, with fuel stored in a "backpack" fashioned of one or more propane tanks.

- Fire balls: Specially constructed juggling balls, either solid ball dipped in fuel and juggled with protective gloves, or ones designed to contain the flame in the center of the ball.
- Wearable fire: Headdresses, hip belts, arm bands, or other garments made typically of metal with Kevlar torches attached. Can be worn while fire dancing.

The variety of available tools took a sharp swing upward in 2000, and as the numbers of dedicated fire tool makers increased, many makers added their own ingenuity to the art and expanded the performance potential even more. Frequently, new tools appear from home tinkering and enter the public domain after a few performances.

5.15.3   Materials and construction

The typical construction of fire performance tools involves a metallic structure with wicking material made from fiberglass, cotton, or Kevlar blended with fiberglass, Nomex, and other poly-aramids. Kevlar-blend wicks are the most common, and are considered standard equipment in modern fire performance. Though most wick suppliers refer to their wick simply as Kevlar, almost no suppliers sell a 100% Kevlar wick, which is both expensive and not particularly absorbent. Most serious contemporary performers avoid cotton and other natural materials because such wicks disintegrate after relatively few uses, and can come apart during use, showering the performer and audience with flaming debris.

A typical poi construction would consist of a single or double-looped handle made of webbing, Kevlar fabric, or leather. This is connected to a swivel and a length of chain or cable. This chain or cable then connects to another swivel, and then to the wick, which is made out of tape wick (a wide, flat webbing made of wick material), or rope wick. The wick material is typically folded or tied to a central core in either a knot or lanyard-type fold.

The chain or cable can be anything from stainless steel wire rope (preferred by some for its low cost, light weight, high strength, and almost invisible profile, but not by others because it tangles easily) to dog chain (preferred by some for its heft and low cost) to industrial ball chain, which is the most common chain for fire performance equipment. Made of nickel-plated steel, stainless steel, or black-oxide brass, ball chain in the #13 to #20 size ranges provides excellent strength, a fluid feel, and great tangle prevention. Since every link on the chain swivels, one can eliminate dedicated swivels from a design, and body wrapping and chain wrapping moves become much easier. Extra cost and a higher weight-to-durability ratio are the biggest downsides to ball chain.

A fire staff typically consists of a long cylindrical section of either aluminum tube (lighter, more suitable for fast-spinning tricks) or wood (heavier, more suited to "contact" moves in which the staff retains contact with the performer throughout the trick; see contact juggling) with a length of wick secured at either end, usually with screws. Wooden-cored staves often have thin sheet metal wrapped around the ends to prevent charring of the wood from the heat—this will have holes drilled through it to allow the wick to be screwed securely into the core. Metal staves generally have a length of wooden dowel inserted into each end; holes are drilled through the metal to allow the wicks' screws to gain firm purchase on the wooden core. A grip of some sort is usually fashioned in the center of the staff to provide a comfortable hand-hold—most commonly leather, or a soft, self-adhesive grip of a type designed for hockey sticks or tennis rackets.

5.15.4   Important factors in equipment construction

Building high quality fire performance equipment involves the balancing of a number of factors to achieve performance suited for the specific intended use by the performer. Even if you are planning on buying prefabricated equipment, understanding the following factors and how they interrelate will allow you to best purchase the right implement.

- Balance: Balance is how the weight is distributed in the implement. It is critical when making staffs, torches, hula hoops, clubs, and swords, as balance will determine the axis around which the implement rotates.
- Weight: Making implements heavier will, up to a certain point, allow you to spin them faster. However weight will also make the implement increasingly unwieldy. Also, heavy implements are more likely to lead to repetitive stress disorder, and cause injuries if you make mistakes. Heavier implements make certain types of contact juggling much easier, and certain high speed manipulation more difficult.
- Wick size: Generally, the more exposed surface area of wick on the prop, the larger the flame. More wick will increase the fuel the implement will hold and if the wick is layered it will increase burn time. The prop will also be heavier and more expensive to construct. The more fuel the prop holds the larger the increase in weight after fuelling.
- Cost: The fourth factor is cost. Frequently, new prop development, and sometimes even building standard designs, requires extra materials and tools that are not readily available. Even dedicated home tinkerers find themselves weighing the cost of purchasing versus the cost and time of build at home.

5.15.5   Fuels

Nearly all modern fire dancing apparatus rely on a liquid fuel held in the wick. There are many choices for fuels, each differing in properties. Individuals select a fuel or a blend of fuels based on safety, cost, availability, and the desirability of various characteristics like color of flame, heat of flame, and solubility. There are also geographic variances in fuels used, based on local availability, pricing, and community perception. For example, American fire spinners commonly use Coleman gas or 50/50 mixes while British fire spinners almost exclusively use paraffin oil (which the Americans call kerosene or jet fuel). Frequently, particularly in areas not fully industrialized, the fuel available is the residue from productions of more refined fuels. Travelling performers can find themselves spinning highly toxic, smoky, or carcinogenic fuels.

- Isoparaffin oil: Some known types include Pegasol 3440 special, Shellsol T, Isopar G. MSDS lists them as naphtha (petroleum), heavy alkylates. Performers seek isoparaffins with low aromatic, benzene, and sulfur contents. These can be odorless, burn clean with little smoke, and are available in a range of flash points. Little is known or used in the United States.
- White gas: Also known as Coleman fuel, naphtha, or petroleum ether. This hot, volatile fuel is popular because it is easy to ignite, burns brightly, evaporates cleanly, and does not leave smoke or residues on wicks and bodies. However, it burns hot and quick, limiting the burn time, and potentially increasing the risk of burns. This is the preferred fuel for performers who do indoor shows in the United States. The fuel is becoming increasingly more difficult to obtain in the United States due to its alternate use in methamphetamine production.
- Kerosene/paraffin oil: This is a popular fuel due to its low cost and long burn times. Kerosene is a generic term that covers a broad range of fuels ranging from gasoline to diesel fuel. It is normally a mixture of hydrocarbons. Almost every maker of kerosene has different purity standards and different flash points. Some home fuel oils are nearly pure paraffins (alkanes and iso-alkanes) whereas others are almost completely benzene and refinery residue.
- Lamp oil: Lamp oil is an oily, non-volatile fuel. Typically sporting the highest flash point of all the petrol distillates in liquid form, lamp oils are the most difficult to light and longest burning fuels. Many products sold as lamp oil contain a limited amount of non-alkane petrol distillates (benzene, etc.), and many have colorings and scent additives that have some toxic potential. Even the purest grades of lamp oil burn quite smoky (though less irritating and toxic), and

thus make it preferred for outdoor use. The soot from burned lamp oil can be difficult to wash out of clothing.

- Alcohol fuels: Usually ethanol, methanol, or isopropyl. Industrial or lab alcohol is usually ethanol with methanol, acetone, or other denaturing agents added. Denatured alcohols can be up to 95% ethanol, or as little as 50%. An MSDS sheet of the mixture will indicate the exact contents.

Note: The flame is blue to orange, depending on methanol content, and fairly dim. However, when mixed with chemicals such as lithium chloride, copper chloride, and boric acid, various colors of flame can be created. Lithium compounds produce pinks, copper compounds produce greens and blues, and boric acid produces green. Other chemicals may produce other colors, and performers often experiment with various choices. Use of chemicals like these may produce some toxic vapors, and have a tendency to destroy wicks. Due to the weak flame, price and toxicity of methanol, it is usually only used for colored flame production and in mixes.

- Biodiesel: Biodiesel is a fuel produced by refinement or transesterification of vegetable oil (used or virgin) using meth oxide composed of methanol and lye. Both KOH, potassium hydroxide, and NaOH, sodium hydroxide, can be used in the process but only one or the other, never both in the same batch. This produces glycerin and methyl esters, aka biodiesel. The fuel is designed for use in diesel vehicles, but is a fairly safe and practical fuel for fire performance. Like kerosene, it is difficult to ignite by itself, and produces a dim, long-lasting flame that may smell a bit like French fries, depending on the source. It is often mixed with white gas to produce an easy-to-ignite, long-burning fuel.

5.15.6   Risk engineering

Metal parts on fire tools have a high heat transfer coefficient and may burn on contact; the wick has a lower coefficient and is less likely to cause burns directly, but can spray or spread fuel. Costumes from non-flammable or flame retardant materials, such as leather or treated cotton, are preferred when employing fire; synthetic materials tend to melt when burned, resulting in severe burns to the wearer.

Fire tools require a safety regime to address the risks of setting fire to the user, bystanders, or the surroundings. Typical elements of such a regimen include a sober, rested, and alert spotter who has access to an ABC dry chemical fire extinguisher for putting out material and fuel fires (water-based extinguishers may spread oil fires), a damp towel or woolen/

duvetyn fire retardant blanket (for extinguishing burning clothes and fire toys), a bucket of water (for the eventuality of out-of-control fires), and plastic wrap (for protecting burns that require hospitalization). Typically, a metal container (located away from the performance area) that can quickly be sealed (so as to be airtight) is used as a fuel dump; with the lid in place, fuel fires may be extinguished.

5.15.7   History

Fire dancing using different techniques is a part of the historic culture of some areas of the world. The oldest practice of fire dancing is Samoa known as Siva Afi and Fire Knife. The fire knife dance has its roots in the ancient Samoan exhibition called "ailao"—the flashy demonstration of a Samoan warrior's battle prowess through artful twirling, throwing and catching, and dancing with a war club while on fire. The "ailao" could be performed with any war club and some colonial accounts confirm that women also performed "ailao" at the head of ceremonial processions, especially daughters of high chiefs. During night dances torches were often twirled and swung about by dancers, although a war club was the usual implement used for "ailao." Ancient Aztecs performed a fire dance dedicated to Xiuhtecuhtli, the god of fire. The Aztec fire dance is performed today for tourists in Mexico. In Bali, the Angel Dance and the Fire Dance, regularly performed for tourists, have origins in ancient rituals. Both the Angel Dance and the Fire Dance originated in a trance ritual called the sang yang, a ritual dance "performed to ward off witches at the time of an epidemic." Also known as the "horse dance" men perform the dance by holding rods representing horses, while leaping around burning coconut husks, and walking through the flames. French Polynesia, Antigua, Cuba, and Saint Lucia are other locations where fire dances are recreated for tourists. The Siddha Jats of the Thar Desert in India perform traditional fire dances as part of the spring festival. Fire dancing is performed to music played on drums and the behr. There are variations of fire dancing; men often perform a dance that involves walking on hot coals, while women perform a dance while balancing flaming tin pots on their heads. Today this ritual is often performed for tourists.

5.15.8   Modern developments in fire performance

During the period from the mid-1990s to the early 2000s, fire dancing grew from a relatively obscure and marginalized native tradition and a talent and skill of the baton twirler or circus artist, to a widespread and almost commonplace occurrence at raves, rock concerts, night clubs, beach parties, camping festivals, cabarets and hotel shows. Many attribute the discipline's rapid growth in popularity to the Burning Man festival, where many thousands were exposed to fire dancing who had never seen or heard of it before. Another powerful force was the rise of Internet chat and bulletin board cultures, which allowed aspiring dancers in isolated areas to communicate with the then-limited pool of skilled performers far outside of their geographic confines.

As the number of fire dancers multiplied exponentially, individual performers and troupes began to experiment with new equipment concepts (i.e., beyond the traditional staff, fire knives, and poi) and with hybrid performance art concepts. The following is an incomplete list of such show varieties, whose categories are general and tend to overlap.

- Traditional fire shows: Traditional shows often incorporate Polynesian costuming and other cultural elements. Many conform to the guidelines or are inspired by the annual World Fire Knife Competition and Samoa Festival.
- Standard modern shows: These usually include performers in tight and perhaps even risqué costumes with elaborate face paint, performing with poi, staffs, and other standard implements. Such shows often include fire breathing techniques as well. Most people think of this type of performance when they think of fire dancing.
- Fire theatre: Such shows are theatrical shows which include fire and fire performance as elements of staged dramatic presentations. Often the fire performance is a small element of the larger show. These shows tend to use more elaborate props and costuming and focus less on technical skill.
- Erotic fire show: Such shows may be seen as simply a normal improvised fire dance but with emphasis on sexually arousing body gyrations, seductive facial expressions, an eroticized musical selection (such as R&B or down tempo music), and minimal clothing of the performer, thus promoting sexual arousal or desire in addition to the expected visual entertainment for an audience. Unlike a fire fetish show, this performance is generally more low-key, slower in tempo, and may be performed by a solo dancer in front of a small and select audience, often a spouse or romantic partner. This performance is considered to be an active and visually exciting form of ritual foreplay. However, this type of show is usually only enticing to a select audience and is generally unpopular by the mainstream community.
- Ritual fire show: Such shows are usually a fusion of pagan or occult ceremony with fire and fire performance. They focus less on technical skill, and more on the use of the fire dancer to highlight the ritual.
- Fire and belly dance: Such shows are a fusion of Middle Eastern belly dancing (raqs sharqi) and combine elements of fire dancing and belly

dancing. Often the dancers use palm torches and fire swords made to resemble scimitars.

- Fire comedy jugglers combine many of the skills of other fire performers but also include juggling, which is rarer in other spinners. The juggler also includes comedy to round out his routines, like lighting his behind on fire.
- Cirque Du Soleil has for the first time incorporated contemporary fire dance techniques in its Zaia production in Macau. Previous Cirque Du Soleil shows "Alegria" and "O" relied on the skills of traditional fire knife artists for fire performances. Recognition of contemporary fire dance and modern prop techniques has previously been very limited in the professional circus community.

Other performance variations continue to emerge as fire dancing becomes more widespread and commonplace.

5.15.9   Physics of firewalking/fire dancing

To discuss the physics of firewalking, we first must delve into the science of heat transfer. Thermal energy is transferred in the following forms:

- Radiation: the transfer of thermal energy via particles carried through electromagnetic waves (think the warming rays of the sun).
- Convection: the transfer of thermal energy through a gas or liquid, in which hot air rises and expands while cold air drops and condenses.
- Conduction: the transfer of thermal energy by direct contact with a heated object.

Apparently the first two methods don't apply to firewalking, seeing as there is very little radiated thermal energy provided by the hot coals and no liquid or gases involved either. This leaves the main method of heat transfer as conduction. There are two schools of thought as to why the feet are not burned by conduction when touching the hot coals in a firewalk, one being more popular than the other.

The first involves the basic principle that wood (including coals) is a bad conductor of heat. The thermal conductivity of charcoal is small, due to its lightweight carbon structure, and that of the feet is only about 4 times more, due to its extra thick skin. This low thermal conductivity can best be explained by the analogy of a cake baking in an oven. If you place your hand in the middle of a hot oven, your hand will not be burned because it is denser than the hot air of the oven. However if you touch any of the metal inside the oven, your hand will be immediately burned since metal is far denser than air. The denser something is, the better conductor of heat it is. The cake inside the oven will feel hot, but can be touched briefly without causing a burn.

Hardwood and charcoal are not only low thermal conductors, but also excellent thermal insulators. For years wood was used for the handles of frying pans and other cookware until heat-resistant plastics were invented. Ash itself is also a poor conductor and a layer of ash atop the hot coals also blocks any thermal energy in the form of radiation from hitting the feet.

In addition to these properties of wood and ash, the fact that the feet come into contact with the coals for such a brief period of time prevents them from being burned. The moment of contact between feet and coals is far too quick to burn the feet, usually half a second or less. Additionally, the entire foot does not make contact with the coals due to the arch of the feet.

The second and less popular school of reason can best be described by an old Boy Scout camping trick to cook an egg. If you take a paper cup, fill it with water, and place it atop hot coals, the water will eventually boil enough to cook an egg without igniting or burning the surrounding paper cup. The reason for this is the water inside the cup keeps the cup below its flash point (lowest temperature at which it ignites). Since the feet contain a large amount of water, some believe this same principle can be applied to firewalking. Many firewalking instructors teach their students to stay relaxed when attempting a walk to keep the blood flowing to the feet, further protecting the skin from potential burns.

Although the science behind it is quite convincing, burns still happen during modern firewalks, either because the coals are too hot, a walker has unusually thin soles, or they are moving too slow. Even though moving too slowly can cause the feet to burn, running causes burns too by pushing the feet deeper into hot coals. The best strategy to firewalk is a light but brisk pace.

Foreign objects (like metals) that are left within the coals can also cause burns. Additionally, if coals are not burned long enough to eliminate all the water from them, they could then be too hot to walk on and also cause burns. Furthermore, not allowing a layer of ash to build atop the coals has caused burns in the past. The longer the stretch of firewalking coals, the more ash should be allowed to build up.

Some believe wetting the feet with water can help first-time firewalkers, but this may also cause the coals to stick to the feet and result in severe burns. The wet foot principle is based upon the Leidenfrost effect, that the foot can be insulated by water vapor when the water surrounding it touches the hot coals. This is the same principle you see in play when you pour drops of water into a hot skillet and watch the droplets scurry around and evaporate slowly. This is because they are traveling on a bed of steam that insulates them from the heat, steam created when exposing a very hot surface to water. Since steam is a gas, it conducts heat less rapidly than water itself.

It's important to note that firewalking is usually only done at night because during the day the coals would simply appear to be a bed of ashes. It is only at night that the heat between the coals is visible, increasing the apparent danger.

Whatever the science behind it may be, it's no doubt that firewalking still takes guts to do! Although it doesn't offer any direct health benefits, over the years firewalking seminars have helped hundreds of people overcome personal fears, struggles, and hurdles in their lives to become the people they have always wanted to be. And some of those who have conquered the hot coals have even moved on to the more insane practice of glass-walking.

chapter six

System risk engineering

Engineering is the art or science of making practical.

Hannah Arendt
Denktagebuch, vol. 2 (1969)

6.1 Edith Clarke and power systems risk engineering

As described in the book titled *What Every Engineer Should Know About Risk Engineering and Management*, power systems risk engineering mitigates risk due to "blackout" and instability, etc. Ms. Edith Clarke pioneered research and engineering development on related subjects for electrical power systems.

Edith Clarke (February 10, 1883–October 29, 1959) was born in Howard County, Maryland. She was the first female electrical engineer, specializing in electrical power system analysis.

In 1919, Edith took a job as a "computer" for General Electric in Schenectady, New York, and in 1921 filed a patent for a "graphical calculator" to be employed in solving electric power transmission line problems. This simple graphical device could solve equations involving electric current, voltage, and impedance in power transmission lines. Also, the device could solve line equations involving hyperbolic functions 10 times faster than previous methods. The patent was granted in 1925.

In General Electric, Ms. Clarke finally achieved her life-long goal—to work as an engineer for the Central Station Engineering Department of General Electric. This made her the first professionally employed female electrical engineer in the United States, and the first woman to be accepted as a full voting member of the American Institute of Electrical Engineers (AIEE, which became IEEE in 1963). She retired in 1945, and became a Fellow of AIEE in 1948; the first woman to be so honored.

Ms. Clarke authored or co-authored 19 technical papers between 1923 and 1951. She was the first woman to present an AIEE paper. This paper, "Steady-State Stability in Transmission Systems," was later published in *AIEE Transactions*. Clarke later earned the AIEE's 1932 Best Regional Paper Prize and the 1941 National Paper Prize.

Additionally, Ms. Clarke authored a two-volume reference text-book, *Circuit Analysis of A-C Power Systems,* based on her lecture notes for General Electric Advanced Engineering Program. This two-volume classic work was published in 1943 and 1950.

In Proceedings of the IEEE, the Institute of Electrical and Electronics Engineers (retrieved October 16, 2012), Dr. James E. Brittain presented a paper titled "From Computer to Electrical Engineer—the Remarkable Career of Edith Clarke." The paper explains why Edith was a pioneer for women in both engineering and computing:

> Edith Clarke's engineering career had as its central theme the development and dissemination of mathematical methods that tended to simplify and reduce the time spent in laborious calculations in solving problems in the design and operation of electrical power systems. She translated what many engineers found to be esoteric mathematical methods into graphs or simpler forms during a time when power systems were becoming more complex and when the initial efforts were being made to develop electromechanical aids to problem solving. As a woman who worked in an environment traditionally dominated by men, she demonstrated effectively that women could perform at least as well as men if given the opportunity. Her outstanding achievements provided an inspiring example for the next generation of women with aspirations to become career engineers.

In 1954, Ms. Clarke received a Lifetime Achievement Award from the Society of Women Engineers. The award cited her contributions to the field in the form of her simplifying charts and her work in system instability.

Edith Clarke retired to her native Maryland in 1956, where she died on October 29, 1959, in Olney, Maryland.

6.2   Encoding the geometry of navigating our stars: Flyby Pluto and beyond

In July 2015, NASA—and the United States—completed the reconnaissance of the planets by exploring the Pluto system with New Horizons (see Figure 6.1).

Figure 6.1 New Horizons: The first mission to the Pluto System. (Courtesy of National Aeronautics and Space Administration.)

Navigation of the New Horizons spacecraft during approach to Pluto and its satellite Charon presented the following new challenges related to the distance from the Earth and sun and the dynamics of two-body motion where the mass ratio results in the barycenter being outside the radius of the primary body:

- Since the Earth is about 30 astronomical units (a.u.) from the spacecraft during the approach to Pluto and Charon, the round trip light time is greater than 8 hours making two-way Doppler tracking difficult. 1 astronomical unit = 92,955,807.3 miles.
- The great distance from the sun also reduces the visibility of Pluto since Pluto receives about 1/900 of the solar radiation as the Earth.
- The two-body motion involves Pluto and Charon moving in elliptic orbits about each other, and the system mass is a simple function of the period and semi-major axis of the orbit.
- The period can be measured to high precision from Earth-based telescope observations and the orbit diameter can be measured to a precision of perhaps 100 km enabling the system mass to be determined within 1%.
- The maximum separation of Pluto and Charon on a star background provides a powerful observation of the total orbit size.
- The fly-by mission is characterizing the geology and atmosphere of Pluto and its large moon Charon. At launch, the spacecraft followed a heliocentric trajectory to a Jupiter fly-by for gravity assist in 2007, and then settling into a long 8-year cruise to the outermost planet.

To navigate further into stars beyond our solar system, a rotating black hole might provide additional gravity assistance, if its spin axis is aligned the right way.

General relativity predicts that a large spinning mass produces frame-dragging—close to the object, space itself is dragged around in the direction of the spin. Although attempts to measure frame dragging about the sun have produced no clear evidence, experiments performed by Gravity Probe B have detected frame-dragging effects caused by Earth.

In general relativity, the gravitational field is encoded in the elliptic geometry of space-time. Much of the conceptual compactness and mathematical elegance of the theory can be traced back to this central idea. The encoding is also directly responsible for the most dramatic ramifications of the theory: the big-bang, black holes, and gravitational waves.

Reaching Pluto, the "third" zone of our solar system—beyond the inner, rocky planets and outer gas giants—has been a space science priority for years because it holds building blocks of our solar system that have been stored in a deep freeze for billions of years. Encoding the geometry of navigating the solar system and beyond, we can fix "The Fault in Our Stars," mitigating risk and uncertainty.

6.3 From radioactive isotopes, space nuclear power, to "Mission to Mars"

Enrico Fermi, Nobel Laureate in physics, was born in Rome on September 29, 1901. Dr. Fermi detected unknown radioactive isotopes through experiments. When Dr. Fermi irradiated heavy atoms with neutrons, these were captured by the nuclei. New isotopes, often radioactive, were formed. The Nobel Prize for physics was awarded to Dr. Fermi for "his demonstrations of the existence of new radioactive elements produced by neutron irradiation, and for his related discovery of nuclear reactions brought about by slow neutrons." Dr. Fermi's achievement led to the discovery of nuclear fission and the production of elements lying beyond what was until then the Periodic Table.

Radioisotope power systems (RPS) now becomes a type of nuclear energy technology that uses heat to generate electric power for operating spacecraft systems and science instruments. That heat is produced by the natural radioactive decay of plutonium-238. Currently, NASA works in partnership with DOE to maintain the capability to produce the multimission radioisotope thermoelectric generator (MMRTG) and to develop higher-efficiency energy conversion technologies, such as more efficient thermoelectric converters as well as Stirling converter technology.

As shown in Figure 6.2, representing a legacy of exploration, the latest RPS to be qualified for flight, called the MMRTG, provides both power and heat for the Mars science laboratory rover.

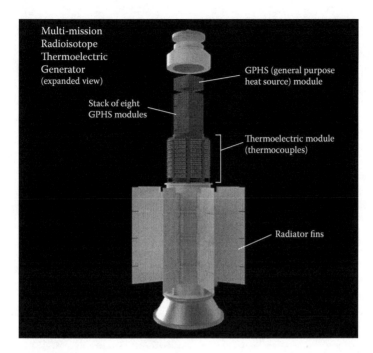

Multi-mission Radioisotope Thermoelectric Generator (expanded view)

GPHS (general purpose heat source) module

Stack of eight GPHS modules

Thermoelectric module (thermocouples)

Radiator fins

Figure 6.2 Major components of the MMRTG, or multi-mission radioisotope thermoelectric generator. (Courtesy of National Aeronautics and Space Administration.)

Published by CRC Press in 2001, the first book on risk engineering, *What Every Engineer Should Know About Risk Engineering and Management*, was motivated by the research work on space nuclear power systems.

The Mars exploration rovers act as robot geologists while they are on the surface of Mars. In some sense, the rovers' parts are similar to what any living creature would need to keep it "alive" and able to explore. The rover has:

- A body: A structure that protects the rovers' "vital organs"
- Brains: Computers to process information
- Temperature controls: Internal heaters, a layer of insulation, and more
- A "neck and head": A mast for the cameras to give the rovers a human-scale view
- Eyes and other "senses": Cameras and instruments that give the rovers information about their environment
- Arm: A way to extend its reach
- Wheels and "legs": Parts for mobility
- Energy: Batteries and solar panels
- Communications: Antennas for "speaking" and "listening"

6.4 *"All the astronauts who landed on the Moon were engineers..."*

The Preface of the book titled *Engineering Robust Designs with Six Sigma* states:

> A few of the book's reviewers work or have worked for the National Aeronautics and Space Administration. This reminds me of these facts: "All the astronauts who landed on the Moon were engineers," and we are following in their footprints when engineering robust products for humankind.

If there has to be one class of professionals the whole world has to be grateful for, then it undoubtedly has to be engineers, for there is no aspect of our lives that has not be made easier or better by these professionals. Engineers are basically the trained professionals who design the processes for and build products, machines, and structures. Engineering is a vast field and there are several categories of engineers each devoted to one area of specialization. Civil engineers plan, design, and construct infrastructure like roads, bridges, buildings, and so on while environmental engineers look at the environmental impact of engineering activities. Mechanical engineers comprise the broadest category of engineers and include manufacturing engineers, aerospace engineers, and acoustical engineers. There may be several types of engineers, but all engineers have one thing in common—they work to make our lives better by problem solving and designing new gadgets, processes, and structures. To be a successful engineer one needs to be highly creative and possess trouble-shooting capabilities. Engineering is a challenging profession as engineers often have to operate under immense pressure, but it can also be extremely rewarding. Browse on to learn about the various famous engineers who have left their marks on the world.

Buzz Aldrin is an American engineer and former astronaut. As the Lunar Module Pilot on Apollo 11, he was one of the first two humans to land on the Moon, and the second person to walk on it. Buzz Aldrin was born on January 20, 1930, in Montclair, New Jersey. His father, a colonel in the U.S. Air Force, encouraged his interest in flight. Aldrin became a fighter pilot and flew in the Korean War. In 1963, he was selected by NASA for the next Gemini mission. In 1969, along with Neil Armstrong, they made history with the Apollo 11 mission when they walked on the moon. As shown in Figure 6.3, Apollo 11 astronaut Buzz Aldrin worked at the deployed Passive Seismic Experiment Package on July 20, 1969. Aldrin later worked in shaping space-faring technology and worked as an author, penning titles like his memoir *Return to Earth*.

Figure 6.3 Apollo 11 astronaut Buzz Aldrin works at the deployed Passive Seismic Experiment Package on July 20, 1969. (Credit: NASA/Neil Armstrong.)

Neil Armstrong happens to be an aerospace engineer and a man who built the very foundation for all the space expeditions that were later enhanced by the oncoming generations. Neil Armstrong was born on August 5, 1930, in Wapakoneta, Ohio. He is of Scot, Irish, and German origin. He is an aerospace engineer and the first person to set foot on the surface of the moon on July 20, 1969, as mission commander of the Apollo 11 moon landing mission. Buzz Aldrin too descended with him on the moon while Michael Collins remained in the command module. Richard Nixon awarded the Presidential Medal of Freedom to Neil Armstrong along with Collins and Aldrin. Neil Armstrong was awarded the Distinguished Boy Scouts Award and during his flight to the moon he sent a greeting to the Scouts saying, "I'd like to say hello to all my fellow scouts and scouters at Farragut State Park in Idaho having a national jamboree there this week; and Apollo 11 would like to send them best wishes."

From the era of cavemen to the era of smart phones, humankind sure saw an incredible journey toward development. It wouldn't be an exaggeration to say that we owe most of it to some of the remarkable engineers of the world. If it weren't for the toil and brains of all the successful engineers of both the past and the present, we wouldn't have been able to get this far as a group of developed nations.

Many of the greatest advancements in history have come about as the direct result of those working as engineers. Engineers provide us with practical solutions for a host of problems, as well as advance practical science and technology. They take theories and ideas, and often turn them into working principles and products that better our lives. From the compound pulley system invented by the great Greek engineer Archimedes, to the tall buildings and air-conditioned comfort we enjoy today, engineers have been at the forefront of our technological advancement.

Engineering shapes critical thinking and problem-solving skills, making engineers a valuable commodity in many different fields. While there have been many notable engineers throughout history, there are some whose inventions and insights have been exceptionally useful. From engineering students tinkering to improve old designs, to the engineers who have discovered sweeping laws that affect the way we view the scientific world. Many people have changed the world, from doctors to politicians, from feminists to philanthropists; but engineers have also played a huge part in the development of the world. Engineers change the way our world works in many different ways. They've revolutionized the way we communicate, get around, and view the world in general.

6.5   European Space Agency's intermediate experimental vehicle splashes down safely

The European Space Agency (ESA) has successfully completed a test run for its Intermediate eXperimental Vehicle (IXV) prototype. The IXV project is paving the way for a future of autonomous reentry vehicles, and the test run demonstrates IXV's ability for safe recovery. The ESA simulated a returning spacecraft by dropping the IXV from a height of nearly 10,000 feet. It gained speed as it fell, and then deployed its parachute, which slowed it down and allowed for a safe splash down just off the coast of Italy.

Similar tests for autonomous reentry vehicles, particularly with regard to the speed and angles affecting water impact, have been previously conducted by CNR-INSEAN of Rome. Tests of supersonic parachutes similar to the one used by the IXV have been conducted in Arizona at the Yuma Proving Ground.

Supersonic parachutes are aerodynamic decelerators used for atmospheric descents. While they can be packed efficiently, they maintain a low mass which maximizes their ability to decelerate even at supersonic (above the speed of sound) or subsonic (below the speed of sound) speeds. Parachutes designed to slow objects falling at supersonic speeds require complex construction, as pressurization and the imbalanced forces of gravity and the parachute's low inertia can cause structural deformations that can compromise the integrity of the parachute.

In an IXV test of May 2014, after the main subsonic parachute was deployed, a series of other actions followed: the parachute's thermal protection covers were cut, and actuators released and jettisoned panels that covered the flotation balloons. Then the parachute prepared for impact with the water and activated beacons allowing it to communicate with a satellite so the craft could be located. All went according to plan except an abnormality in the balloon inflation, which will be analyzed. Researchers recovered the IXV prototype and are currently inspecting it for indications of future improvements.

The IXV will undergo further tests on its hypersonic and supersonic systems next year, when it is sent into a suborbital path on the Vega launcher and will then reenter Earth's atmosphere, eventually making another water landing.

The IXV represents another phase of the development and implementation of autonomous vehicles in space exploration. As of yet, no reports indicate the ability of such crafts to do push-ups on the Moon, however.

6.6   Systems risk engineering: Shall we vote electronically in November's election?

Approaching a crossroads in Michigan, a researcher or hacker may be just hacking into the traffic lights.

For November's 2014 midterm elections, risk engineering of electronic voting systems presented a major technical challenge. According to *USA Today*, "After a decade of use, a generation of electronic voting equipment is about to wear out and will cost tens of millions to replace."

The unsolved problems included the ability of malicious actors to

- Intercept Internet communications,
- Log in as someone else, and
- Hack into servers to rewrite or corrupt code.

While these are also big problems in e-commerce, if a hacker steals money, the theft can soon be discovered. A bank or store can decide whether any losses are an acceptable cost of doing business. Computer scientists and fair-election advocates have warned for years that potential software malfunctions are possible threats to the integrity of elections that use electronic voting systems. Technological flaws can be exploited to gain access to or privilege within a computer system. For example, software vulnerabilities can be exploited to gain administrator control. Vulnerabilities can often be due to the software not being up to date. In rare cases, hackers exploit previously unknown (and therefore unprotected) software flaws, so called zero-day attacks.

Take direct recording electronic (DRE) systems as an example. Using one of three basic interfaces (pushbutton, touchscreen, or dial), DRE systems enable voters to record their votes directly into computer memory. The voter's choices are stored in DREs via a memory cartridge, diskette, or smart card and added to the choices of all other voters. An alphabetic keyboard is typically provided with the entry device to allow for the possibility of write-in votes, though with older models this is still done manually.

Most DRE systems have been introduced since federal legislation was enacted in 2002 that allocated $4 billion toward modernizing the voting process. There are several makes and models, and user interfaces vary. All the DRE systems rely on computers to register and store votes. This makes them vulnerable to software bugs that could under-count or over-count votes. The DRE systems that don't produce paper records are considered the riskiest because a malfunction in such machines could be impossible to detect, much less fix.

The swing states are more vulnerable to glitches that could tip the election. In these states, midterm elections will likely be extremely close, magnifying the potential impact of vote-counting errors.

Shall we vote electronically in November's Election Day? Remember, fundamental network security problems remain to be solved. Effective approaches to cyber security integrate technological measures with robust processes and reliable products.

6.7   Risk engineering of the first transatlantic telegraph cable

Through risk engineering, engineers have successfully achieved reliability growth of the transatlantic telegraph cabling systems.

In the mid-1850s, telegraph cables stretched across much of the United States and the United Kingdom, allowing people within those countries to quickly communicate with one another. If someone wanted to send a message from one of these countries to the other, however, they had to do so the old-fashioned way—by boat, a process that usually took about two weeks. Clearly a telegraph line connecting the United States and the United Kingdom would benefit both nations, but the ocean dividing them seemed too difficult to overcome. That is, to everyone but a young, enthusiastic New Yorker named Cyrus Field, who had made his fortune in paper manufacturing.

The cable shown in Figure 6.4 had been manufactured in two sections and was too large to be carried by a single ship.

On their first attempt in 1857, the ships left Ireland together, laying one ship's cable, then splicing and laying the second ship's load. This failed when the cable snapped and was lost.

On the second attempt, they began laying the line from a midpoint in the Atlantic. After splicing the ends of their cables together, they took off in opposite directions. The line snapped right away, but workers were able

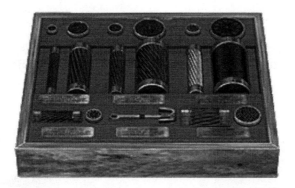

Figure 6.4 Transatlantic telegraph cable—1858. (Courtesy of National High Magnetic Field Laboratory.)

to reconnect the ends and keep going for a short period. After another break, another splice, and yet another break, the ships returned to Ireland in defeat. Because the line broke so quickly, however, the team found they had enough cable to try again.

On the third attempt, the engineering team achieved the project goal; both boats made port at their final destinations on August 5, 1858. Cyrus Field's dream was finally realized and the news of the transatlantic line was greeted enthusiastically on both sides of the ocean.

Queen Victoria sent the first public message across the cable to President Buchanan. The first official message sent along the cable on August 16 read, "Europe and America are united by telegraphic communication. Glory to God in the highest, on earth peace, goodwill to men."

Because of design problems, the process was quite slow (more than 16 hours to send and receive less than 100 words), though considerably faster than sending a message by sea. On August 5, 1858, the first transatlantic telegraph cable was completed.

The problems of the cable voyages, however, were to continue. Sheer determination had got the cable across the ocean, but there had been faults in its manufacture and it had been damaged by the cable laying machinery. The final blow came when the engineer Edward Orange Wildman Whitehouse insisted on using high voltage instruments which further damaged the cable. It stopped working on October 20, 1858.

Just weeks after its completion, the transatlantic cable stopped functioning. The world was appalled and a scandal erupted. A Committee of Inquiry launched an investigation in the United Kingdom, and Whitehouse, the project's chief electrician, received the brunt of the blame. Whitehouse's cable design was faulty. The copper wire core of the line was too thin and he used massive induction coils to send extremely powerful electric currents in hopes that it would speed message transmission. Other experts had questioned his approach, but Whitehouse ignored them. The strength of the currents Whitehouse used was ultimately blamed for over-burdening the cable and causing a breakdown of its insulation (primarily consisting of layers of gutta-percha).

Many years passed by before another cable stretched across the Atlantic. When it was finally in placed in 1866, it formed a permanent link between the United States and the United Kingdom.

6.8 A nation's strength: How a building stands up

In a beautiful poem titled "A Nation's Strength," Ralph Waldo Emerson reveals the secret of a nation's success. The success of a nation lies with the power of its women and men, and not with jewels and gold. Like discussed by the book titled *What Every Engineer Should Know About Risk Engineering and Management* on "How Buildings Stand Up," Emerson compares a nation to a building.

A Nation's Strength

What makes a nation's pillars high
And it's foundations strong?
What makes it mighty to defy
The foes that round it throng?
It is not gold. Its kingdoms grand
Go down in battle shock;
Its shafts are laid on sinking sand,
Not on abiding rock.
Is it the sword? Ask the red dust
Of empires passed away;
The blood has turned their stones to rust,
Their glory to decay.
And is it pride? Ah, that bright crown
Has seemed to nations sweet;
But God has struck its luster down
In ashes at his feet.
Not gold but only men can make
A people great and strong;
Men who for truth and honor's sake
Stand fast and suffer long.
Brave men who work while others sleep,
Who dare while others fly...
They build a nation's pillars deep
And lift them to the sky.

Ralph Waldo Emerson (May 25, 1803–April 27, 1882)

This is a beautiful poem consisting of six stanzas. However, the poem has a very deep message in it. The poet has revealed the secret of how we can make a nation strong. Emerson is of the view that it is the people of a nation not its wealth, gold, jewels, and other natural resources that make it strong. The people who make their nation strong are great. They are not selfish. They work for the prosperity of the nation. They fight for truth and for the sake of honor of their nation. They believe that whatever they do publically or individually, their actions are representatives of the character, honor, and respect of their nation. They do not do anything that can dampen the image and reputation of their nation. To make their nation great and strong, they don't need wealth and jewels. It is their power and determination that makes it strong. The author uses vivid imagery and metaphors to convince readers that the strength of a nation is its people, rather than its wealth or military prowess.

Emerson begins the poem with a metaphor, comparing a nation to a building; "pillars" creates an image of a massive municipal structure (1).

What keeps this foundation from cracking? Not wealth—the foundation is "laid in sinking sand and not on abiding rock" (7–8).

Wealth comes and goes; to be able to withstand the struggles of time one must have a more substantial foundation than money. He asks a rhetorical question: "Is it the sword?" (9).

Does strength in battle equate to a strong nation? The author uses an image of this building that represents a nation, its stone is stained with blood, reduced to rubble from its endless wars (11).

It's "glory [changes] to decay" (12).

History looks on the nation's battles as slaughter rather than glorious conquest. Again Emerson asks a rhetorical question to grab the reader's attention: "And is it pride?" (13).

Emerson compares pride to a "bright crown" which connotes an arrogant king (13).

In the end, it is not gold, or pride, or battle prowess that makes a nation strong; it's those "brave men who work while others sleep [and] dare while others fly" (21–22).

These are the great people who are the actual heroes of the world. If a nation has such people in it, then no one can stop this nation from making progress and making itself strong. We have a rich history of the nations who were small but they had great people in them and those people brought their nations to the heights of progress and prosperity. In the last stanza, Emerson describes the characteristics of brave people. He says that these people do not sleep. They work while others sleep because they don't have time to waste. They are the daring people who face all the challenges and problems of life manfully. They are not cowards like others who step back from the challenges. They build the pillars of a nation very strong. They lay the foundations of their nation very deep. And then they lift the nation to the heights of skies.

It is the hearts and minds of those who toil every day to build our hospitals, teach our children, engineer our dreams, and defend our sweet homes.

With the strength of the people, a nation has a solid foundation and the building stands up.

6.9   Risk engineering and political science

As a giant in the Brobdingnagian realm of world literature, as an imaginary candidate of governor for the "Empire State" (New York State), would Mark Twain, Samuel Langhorne Clemens (November 30, 1835–April 21, 1910), also be a genius in the political science, engineering the risk with political power?

Yes, Mr. Twain has authored a book titled *The Curious Republic of Gondour* (1875), where the government system would be characterized by:

- All the citizens have at least one vote on their Election Day.
- Additional votes could be secured based on education, and so on.
- These additional votes would be provided by the state for free.
- Moreover, no one was accepted to any public office without passing stringent competitive examinations.

Thus, the risk with political power would be engineered with high statistical confidence.

6.10   The lake of no name and thinking
of Chinese poetry

"Do you really believe
he
belongs to
the hill
that provides
him
a home
sweet
and
quiet?"

"I believe
he

belongs to
the lake
of
no name
but always
remembers
his dream ..."

The Lake of No Name is no doubt the most famous lake in all Chinese university campuses. The lake, the tower, and the library (of significance to industrial design engineering in China) are symbols of one of the most prestigious schools in China. For a university that is known for the cradle of Mr.De (democracy) and Mr.Sci (science), this is the spiritual homeland for the conscience of Chinese intellects.

Choice Poetry of Tang and Song is a comprehensive poetry anthology of the Tang and Song dynasties authorized by Emperor Qianlong in the early period of his reign. The compiling purpose of the collection is chiefly characterized by its esteem for the poems of Tang and particular esteem for Du Fu, its emphasis on the function of political enlightenment of poetry, and much sensitivity to the content about national struggle. It is artistically characterized; by careful and accurate selection as well as the valuable poetic viewpoints reflected in the selected items and some of the comments.

In the vein of early Chinese literature, the concept of "significance" serves to epitomize, as "signification" does in modern semiotics, the indeterminate relationship between literary configuration of the world (form) and the spiritual freedom of man (meaning). Starting with a survey of Western ideas on the subject, this research looks into the Chinese equivalent, topically analyzed within five component discourses:

1. Myth, especially the mythological accounts centering on the spiritual freedom of a primordial sage—the Chinese archetype of man.
2. The orthodox literary discourse of poetry modelled on the Classic of Poetry, the early interpretation of which came to formulate fundamental rules for poetic thinking.
3. The preliminary literary discourse of narrative talks, argued here to be the precursor of petty talks, the Chinese designation for prose fiction.
4. The metalinguistic notion of ming, or Names, under which the relationship between language and meaning was fruitfully debated in the classical era.
5. A quintessential literary discourse not only referring to prosaic writing in a rhetorical and ornate style, but gradually to the supreme art of letters.

We can reach an eclectic synthesis of literary theories, seeing them as successive attempts to normalize the relationship between form and meaning, with the loci of significance varying with the flux of poetics and hermeneutics. The distinction between Eastern and Western literary thought, however, partly lies in that, whereas the epistemological pendulum has kept swinging between mimesis (i.e., a real and thus reliable representation) and poiesis (i.e., a false and artificial creation) in the West, Chinese literary minds have long been comfortable with the spirit of transformation, which, in turn, transforms industrial designs in China.

6.11   *Baseball reliability engineering*

Baseball, a bat-and-ball game played between two teams of nine players each who take turns batting and fielding, has been celebrated as "America's National Pastime." Here is a summary of the "fact sheet" of baseball engineering:

Alexander Cartwright, a former player, is credited with formulating the first set of rules.

The first game of record was between the New York Knickerbockers and another New York team, played on June 19, 1846, at the Elysian Field in Hoboken, New Jersey. As shown by Figure 6.5, Spalding's Base Ball Guides, 1889–1939 comprises a historic selection of Spalding's Official Base Ball Guide and the Official Indoor Base Ball Guide.

Figure 6.5 Page 1 of Spalding's official baseball guide, 1913. (Courtesy of Library of Congress.)

Baseball reliability scorecard: A player scores a run when he safely touches first, second, third, and home or his team makes three outs.

Baseball reliability prediction: Forecast a team's future winning percentages given its runs scored and runs allowed.

Baseball reliability modeling: Regression techniques described by *What Every Engineer Should Know About Decision Making Under Uncertainty*.

6.12   *Entropy among the forest of fault trees: When two roads are "really about the same"*

The Road Not Taken

Two roads diverged in a yellow wood,
And sorry I could not travel both
And be one traveler, long I stood
And looked down one as far as I could
To where it bent in the undergrowth;

Then took the other, as just as fair,
And having perhaps the better claim,
Because it was grassy and wanted wear;
Though as for that the passing there
Had worn them really about the same,

And both that morning equally lay
In leaves no step had trodden black.
Oh, I kept the first for another day!
Yet knowing how way leads on to way,
I doubted if I should ever come back.

I shall be telling this with a sigh
Somewhere ages and ages hence:
Two roads diverged in a wood, and I—
I took the one less traveled by,
And that has made all the difference.

Robert Frost

The traveler comes upon a fork in the road while walking through a yellow wood. He considers both roads and concludes that each one is equally well-traveled and appealing. The traveler describes how the two

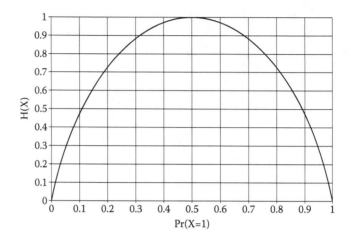

Figure 6.6 Entropy among the forest of fault trees: When two roads are "really about the same."

roads are basically the same. They "equally lay" and were "just as fair" as each other and were even "worn... really about the same."

Why does the traveler feel uncertain about which road to choose? As illustrated by Figure 6.6, looking into the forest of fault trees, the following based on Shannon entropy is revealed:

> A fault tree diagnosis methodology which can locate the actual minimum cut set (MCS) in the system in a minimum number of inspections is presented. An entropy function is defined to estimate the information uncertainty at a stage of diagnosis and is chosen as an objective function to be minimized. Inspection which can provide maximal information should be chosen because it can minimize the information uncertainty and will, on average, lead to the discovery of the actual MCS in a minimum number of subsequent inspections. The result reveals that, contrary to what is suggested by traditional diagnosis methodology based on probabilistic importance, inspection on a basic event whose Fussell-Vesely importance is nearest to 0·5 best distinguishes the MCSs.

Here, we can apply Shannon entropy to minimize the information entropy, and "that has made all the difference" among the forest of fault trees.

Diagnostics is the art or practice of diagnosis (reactive). Prognostics is an engineering discipline focused on predicting the time at which a system or a component will no longer perform its intended function with high statistical confidence (proactive).

Big data is a popular term used to describe the exponential growth and availability of data, both structured and unstructured in terms of the following factors:

- Variability (Six Sigma)
- Volume, velocity, and variety (Lean manufacturing)
- Uncertainty (decision making)
- System diagnosis
- Risk engineering (remaining useful life)
- Environmental impact
- Business communication (risk communication)
- Cost
- Time-dependent system dynamics (phased-mission)

Shannon entropy facilitates the sensing/monitoring, searching/diagnosing, and machine learning/prognostics at the age of big data.

6.13   Space system risk engineering and 3D printing

The International Space Station represents a great endeavor for space system risk engineering (Figure 6.7). The International Space Station's 3D printer has manufactured the first 3D printed object in space, paving the way to future long-term space system risk engineering. The object, a printhead faceplate, is engraved with the names of the organizations that collaborated on this space station technology demonstration: NASA and Made In Space, Inc., the space manufacturing company that worked with NASA to design, build, and test the 3D printer.

An image of the printer, with the Microgravity Science Glovebox Engineering Unit in the background, was taken in April 2014 during flight certification and acceptance testing at NASA's Marshall Space Flight Center in Huntsville, Alabama, prior to its launch to the station aboard a SpaceX commercial resupply mission. The first objects built in space were returned to Earth in 2015 for detailed analysis and comparison to the identical ground control samples made on the flight printer prior to launch. The goal of this analysis is to verify that the 3D printing process works the same in microgravity as it does on Earth.

The printer works by extruding heated plastic, which then builds layer upon layer to create three-dimensional objects. Testing this on the

Figure 6.7 International Space Station. (Courtesy of National Aeronautics and Space Administration.)

station is the first step toward creating a working "machine shop" in space. This capability may decrease cost and risk on the station, which will be critical when space explorers venture far from Earth and will create an on-demand supply chain for needed tools and parts. Long-term missions would benefit greatly from onboard manufacturing capabilities. Data and experience gathered in this demonstration will improve future 3D manufacturing technology and equipment for the space program, allowing a greater degree of autonomy and flexibility for astronauts.

The 3D Printing In Zero-G Technology Demonstration (3D Printing In Zero-G) experiment demonstrates that a 3D printer works normally in space. In general, a 3D printer extrudes streams of heated plastic, metal, or other material, building layer on top of layer to create three-dimensional objects. Testing a 3D printer using relatively low-temperature plastic feedstock on the International Space Station is the first step toward establishing an on-demand machine shop in space, a critical enabling component for deep-space crewed missions and in-space manufacturing.

Three-dimensional printing offers a fast and inexpensive way to manufacture parts on-site and on-demand, a huge benefit to long-term missions with restrictions on weight and room for cargo. After testing of hardware for 3D printing on parabolic flights from Earth resulted in parts similar to those made on the ground, the next step was testing aboard the Space Station. The test included printing items designed by students and results showed that 3D printers work normally in space. This work will contribute to establishing on-demand manufacturing on long space missions and improving 3D printing methods on the ground.

In addition to safely integrating into the Microgravity Science Glovebox (MSG), the 3D print requirements include the production of a 3D multi-layer object that generates data (operational parameters, dimensional control, and mechanical properties) to enhance understanding of the 3D printing process in space. Thus, some of the prints were selected to provide information on the tensile, flexure, compressional, and torque strength of the printed materials and objects. Coupons to demonstrate tensile, flexure, and compressional strength were chosen from the American Society for Testing and Materials (ASTM) standards. Multiple copies of these coupons are planned for printing to obtain knowledge of strength variance and the implications of feedstock age. Each printed part is compared to a duplicate part printed on Earth. These parts are compared in dimensions, layer thickness, layer adhesion, relative strength, and relative flexibility. Data obtained in the comparison of Earth- and space-based printing are used to refine Earth-based 3D printing technologies for terrestrial and space-based applications.

6.14   Decision-making under uncertainty: Eisenhower decides on D-Day

Have you ever had to make a tough decision under great uncertainty? In the spring of 1944, General Dwight D. Eisenhower, the Allied Supreme Commander in Europe, had to make one of the most important decisions of World War II and time was quickly running out. Hundreds of thousands of Allied troops, sailors, and airmen awaited his orders to begin Operation Overlord, the invasion of Normandy. However, the Allied planners knew they could not control the weather for D-Day, originally scheduled for June 5.

6.14.1   Weather forecast

Late on the evening of June 2, 1944, Eisenhower, his top generals, and British Prime Minister Winston Churchill met to review the weather forecast. The news was not good—D-Day, June 5, promised cloudy skies, rain, and heavy seas. Eisenhower decided to wait another day to see whether the forecast might improve. Less than 24 hours before the scheduled invasion Eisenhower gathered his advisers again. The forecast indicated that the rain would stop and there would be breaks in the clouds by mid-afternoon on June 5. What do you think Eisenhower did?

6.14.2   Eisenhower's decision

Eisenhower decided to change the date for D-Day to June 6. He knew that the tides would not favor an invasion again for nearly two weeks, long enough for the Germans to possibly learn of the Allies' plan. Eisenhower gave the order and set in motion the largest amphibious invasion in world history; an armada of over 4000 warships, nearly 10,000 aircraft, and about 160,000 invasion troops (see Figure 6.8).

6.14.3   Great victory

The hard fought invasion was a success—Eisenhower had won his great gamble with the weather. Within 2 months, Allied forces broke out from their Normandy beachheads and began the long heroic struggle to liberate Europe from Nazi tyranny.

Figure 6.8 Decision-making under uncertainty: Eisenhower decides on D-Day. (Courtesy of www.army.mil.)

6.15   "Home is best": Engineers from the Moon to the Earth

The book titled *Engineering Robust Designs with Six Sigma* includes the following:

> A few of the book's reviewers work or have worked for the National Aeronautics and Space Administration. This reminds me of these facts: "All the astronauts who landed on the Moon were engineers," and we are following in their footprints when engineering robust products for humankind.

However, the most exciting moment was recorded by NASA's photo shown in Figure 6.9. After the most famous voyage of modern times, two great engineers were approaching their sweet home.

The most famous voyage of modern times was initiated by the inspiring speech, "before this decade is out, of landing a man on the Moon and returning him safely to the Earth," which requires a systematical process for "robust design."

Figure 6.9 "Home is best": Engineers from the Moon to the Earth. (Courtesy of National Aeronautics and Space Administration.)

6.15.1   Engineering a robust kite with Six Sigma

Imagine that you are standing on a beach flying a kite in the wind; you feel a strong force from the kite in the rope and notice that the kite flies fast, way faster than the wind is blowing. The road map for engineering a robust kite with Six Sigma is summarized as follows:

- Critical to Quality (CTQs) represents the product or service characteristics that are defined by the customer (internal or external).
- CTQs may include the upper- and lower-specification limits or any other factors related to them.
- A CTQ characteristic—what the customer expects of a product— usually must be translated from a qualitative customer statement into an actionable, quantitative business specification.
- It is up to engineers to convert CTQs into measurable terms using Six Sigma tools.

In robust parameter design, which has received considerable attention from academia and industry, one optimally selects the levels of a set of controllable variables (a.k.a. inputs) in order to minimize variability in an output response variable, while keeping the mean of the response variable close to a target. The component of the response variability that can be affected by adjusting the inputs is typically assumed to be due to a set of uncontrollable (a.k.a. noise) variables. Hence, minimizing response variability amounts to choosing the inputs so that the output response is robust or insensitive to variations in the noise variables.

6.15.2   Robust design—interplay of control and noise factors

Industrial robust design methods rely on empirical process models that relate an output response variable to a set of controllable input variables and a set of uncontrollable noise variables. Robust design reflects an interplay of control and noise factors with a methodology described as follows:

- Control energy transformation for each CTQ characteristic
- Determine control and noise factors
- Assign control factors to the inner array

As shown in Figure 6.10, guitar design is a type of industrial design where the looks and efficiency of the shape as well as the acoustical aspects of the guitar are important factors. The guitar is one of the oldest known instruments, tracing its roots back to the oud of Ancient Mesopotamia. How the guitar works is relatively simple: a set of strings is stretched and tensed over the body of the guitar. The body of the guitar consists of a

Figure 6.10 Robust design—interplay of control and noise factors.

hollow cavity. When one of these strings is plucked, the string vibrates at a certain frequency, producing sound. The cavity resonates and amplifies certain frequencies, producing the signature guitar sound. Other stringed instruments work in a similar fashion. The cavity of the instrument and where the strings of the instrument are generally plucked determine the harmonics and identifiable sound of stringed instruments. Over the centuries, many models of the guitar have been created: materials used for the base and strings have changed and there exist many variations in the number of strings among guitar models.

In the past many guitars have been designed with all kinds of odd shapes as well as very practical and convenient solutions to improve the usability of the object. CAE-based optimization has a long tradition in engineering. The goal of optimization is often the reduction of material consumption while pushing the design performance to the boundaries of allowable stresses, deformations, or other critical design responses. At the same time, safety margins are asked to be reduced and products should be cost-efficient and not over-engineered. Of course, a product should not only be optimal under one possible set of parameter realizations. It also has to function with sufficient reliability under scattering environmental conditions. In the virtual world we can prove that, for example, with a stochastic analysis, which leads to CAE-based robustness evaluation. If CAE-based optimization and robustness evaluation is combined, we are entering the area of robust design optimization (RDO) which is also called design for Six Sigma (DFSS) or just robust design (RD).

The main idea behind that methodology is that uncertainties are considered in the design process. These uncertainties may have different sources like, in the loading conditions, tolerances of the geometrical dimensions and material properties caused by production or deterioration.

Some of these uncertainties may have a significant impact on the design performance which has to be considered in the design optimization procedure. Starting with an initial sensitivity analysis, the important design parameters can be identified and the optimization task can be significantly simplified. Taking into account uncertainties, the optimization task becomes more challenging. Instead of deterministic response values, uncertain model responses need to be analyzed. For a successful implementation, this analysis requires the estimation of the probabilities of rare events. With the help of a variance-based and reliability-based robustness evaluation, the required safety level can be implied in the optimization process and verified for the final design.

Successful integration of RDO strategies into CAE-based virtual product development cycles needs an RDO strategy which is in balance with the available knowledge about uncertainties of scattering variables, with available criteria to reliably quantify robustness or safety of designs as well as with the dimensionality and non-linearity of the RDO task.

For definition of successful objectives and criteria for robust designs, sensitivity analysis in the design space of optimization as well as in the space of scattering variables is very helpful. Any kind of design space defined by a handful or hundreds of optimization parameters is a valid space to optimize the design. For real world RDO applications we have to expect that at least in the robustness space we have to start with a large number of potentially important scattering variables. In contrast to the design space of optimization, the variable reduction in robustness space starting from all possible influencing variables is only possible with knowhow about the unimportance of scattering variables.

If the RDO task is defined with appropriate robustness measures and safety distances, multiple optimization strategies can be performed successfully to drive the design in the direction of being optimal and robust. If a design evaluation needs significant time, the balance between the number of CAE design runs and the accuracy of robustness measures is a challenge for all RDO strategies, iterative or simultaneous. Then all of them try to minimize the number of design evaluations to estimate the robustness measures. If small failure probabilities (like smaller than 1 out of 100) need to be proven, algorithms of reliability analysis have to be applied, at least at the end of an RDO process to prove the optimal design.

6.16   *Design: Where engineering meets art*

Design is where engineering meets art
where engineers and artists gather
to inspire us
refresh us
and connect us …

Robust design is an experimental method to achieve product and process quality through designing in insensitivity to noise based on statistical principles. Parameter design is a principle that emphasizes choosing the proper levels for the controllable factors in a process for manufacturing products. When said to be optimal, the implication is that the design has achieved most of the target values set out by the quality measure before proceeding to a tolerance design. In an industrial setting, totally removing noise factors can be very expensive. Through a parameter design, engineers try to reduce the variation around the target by adjusting control factors' levels rather than by eliminating noise factors. By exploiting nonlinearity of products or systems, parameter design achieves robustness, measured by a signal-to-noise ratio (SNR, S/N), at a minimum cost. Orthogonal arrays are used to collect dependable information about control factors with a small number of experiments.

In early 1968, the U.S. Air Force ordered three X-25 type aircraft to test methods of improving the odds of a downed flyer's escape. At the time, the USAF was suffering heavy losses in the Rolling Thunder air campaign over North Vietnam.

The unpowered Bensen X-25 Discretionary Descent Vehicle (also called a "Gyroglider") theoretically could be stowed in an aircraft, ejected with the pilot and deployed during descent. Its rotary wings would be brought up to speed as it fell, and the pilot would fly it as an autogyro to a safer landing area.

As shown by Figure 6.11, the X-25A gyrocopter on display at the National Museum of the USAF represented a more advanced concept

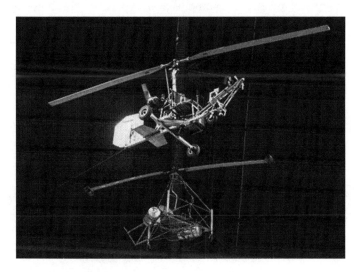

Figure 6.11 Bensen X-25A gyrocopter (front) and McDonnell XH-20 Little Henry (rear) in the Research & Development Gallery at the National Museum of the U.S. Air Force on December 28, 2015. (Courtesy of the National Museum of the USAF.)

with a limited "fly-away" capability. Though similar to the X-25, the X-25A had a more robust structure, and it was powered by a small engine. The two-seat X-25B was originally used as an unpowered, towed trainer, but it was later fitted with an engine.

Tests proved that pilots could be quickly and easily trained to fly the X-25. Even so, with the air war in Vietnam winding down—and doubts about its operational feasibility—the X-25 program ended. The X-25A was delivered to the museum in 1969. For the case study discussed by the book titled *Engineering Robust Design with Six Sigma,* a Six Sigma design team worked to develop a robust gyrocopter—a simple child's toy, made of paper, that presents some interesting and challenging design problems. The idea of designing for robustness, then tuning to target performance, is critical to robust design.

6.17   Edge of Tomorrow, *design for survivability, and stress-strength interference*

Edge of Tomorrow is a 2014 American science fiction action film, which takes place in a future where Earth is invaded by an alien race. When Earth falls under attack from invincible aliens, no military unit in the world is able to beat them. Major William Cage, an officer who has never seen combat, is assigned to a suicide mission. Killed within moments, Cage finds himself thrown into a time loop, in which he relives the same brutal fight—and his death—over and over again. However, Cage's fighting skills improve with each encore, bringing him and a comrade ever closer to defeating the aliens. Lessons learned from the movie *Edge of Tomorrow*: for robust engineering, stress-strength interference enables us to assess survivability, the ability to remain alive or continue to exist.

As shown in Figure 6.12, probabilistic evaluation of stress-strength interference facilitates design for survivability during combat and battles.

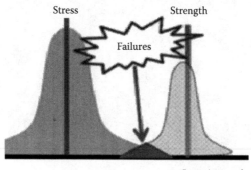

Figure 6.12 Design for survivability, and stress-strength interference.

In structural reliability engineering, one often encounters situations where the strength of a structure is influenced by the stress, but the stress is irrelevant to the strength. This phenomenon can be called a unilateral dependency of strength on stress. The stress–strength interference method is one of the oldest methods of structural reliability analysis. Although more powerful methods of reliability analysis such as the first-order/second-order reliability methods and simulation techniques (which are applicable to a broader class of problems and with less restrictive assumptions) are now available, the stress–strength interference method continues to be a popular method of reliability analysis among practicing engineers in many industries. The attractiveness of the method lies in its simplicity, ease, and economy.

Major William Cage's "Live Die Repeat" illustrates a process of strength increase, establishing a leading edge. Such a leading edge enables Major Cage to prevail eventually.

6.18   *Maze runner versus labyrinth seal*

The Maze Runner, a 2014 American science fiction action thriller film, follows 16-year-old Thomas who awakens in a rusty elevator with no memory of who he is. Thomas has been delivered to the middle of an intricate maze, along with a slew of other boys, who have been trying to find their way out of the ever-changing labyrinth, a maze that will require him to join forces with fellow maze runners for a shot at escape.

By chance, a non-contacting labyrinth seal is introduced for "Applying Axiomatic Design to Design for Six Sigma of Innovative Integrated Systems" (Proceedings of ICAD2002, Second International Conference on Axiomatic Design, 2002, Cambridge, MA, June 10 and 11), which contains the following:

> Imagine two examples of any element from two dra-
> matically different parts of the engineering design
> > Seal versus leakage
> > Tight contact versus labyrinth/maze

As shown in Figure 6.13, the referenced preceding paper includes the following:

> As an integrated part of the innovative lubrication
> and sealing system, sealing is established via the
> high flow resistance. This is created by the turbu-
> lence under high-speed rotating conditions of the
> large crankshaft. During engine start-up and low-
> speed conditions, the oil flows through the grooves,

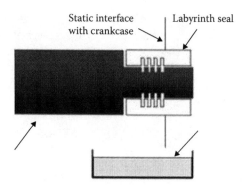

Figure 6.13 A labyrinth seal.

> back to the crankcase oil pan due to vacuum. When comparing with other design alternatives, such as a teethed crankshaft, a grooved crankshaft has excellent capability for manufacture and assembly, as the one-piece labyrinth housing is assembled to the crankshaft. Extensive modeling and testing of the manufacturing process capability, has demonstrated the excellent capability for manufacture and assembly...

Furthermore, TRIZ can be implemented by axiomatic design, using the following two methodologies:

- Decoupling conflicting features
- Simplify information flow

6.19   Axiomatic design: Imagination inspires us to create

Decoupling conflicting features in axiomatic design, which is built upon the following principle in information science: Simply information flow.

In "Flying to Venus: TRIZ for Global Business Relationships," Dr. Wang extends his perspective to

- Global business relationships
- Risk communication

- Aesthetic engineering
- Environmental art
- Conflicts/contradictions

Shown partially by Figure 6.14 of *Flying to Venus*, Sandro Botticelli's painting *Birth of Venus* confronts us with our responsibility in the sphere of global business relationships. This involves defining, recognizing, and fulfilling specific commitments in our relationships with global business partners. These global business relationships highlight the limits and crisis areas for risk communication. The same situation also applies in the sphere of artistic and aesthetic engineering, our relationship with the environment and universe.

According to one perspective from TRIZ (theory of inventive problem solving), the *Birth of Venus* could facilitate our awareness of the multidimensional nature of global business relationships, in particular on the types of potential conflicts. Beyond our awareness of the business relationships with global partners, contradictions are buried subconsciously. These contradictions continue to have a decisive effect on the outcome of our global business relationships.

The picture is a human-made waterfall in a desert-like area.

Here waterfall and desert present conflicts. While discovering inspiration in this visual combination of contradictions in nature may be

Figure 6.14 Flying to Venus, part of Sandro Botticelli's painting *Birth of Venus*. (Courtesy of ItalianRenaissance.org, http://www.italianrenaissance.org/botticelli -birth-of-venus/.)

difficult, imagination empowers us to be creative. Imagine two examples of any element from two dramatically different parts of the engineering design:

- Seal versus leakage
- Tight contact versus labyrinth/maze

Now let's try to combine them into a single workable composition. We may "achieve robust designs with Six Sigma."

chapter seven

Contingency planning, logistics, and Lean manufacturing

Rolling out the storm

> When solving problems, dig at the roots instead of
> just hacking at the leaves.
>
> **Anthony J. D'Angelo**
> *The College Blue Book*

As the basis of industrial design engineering, the process of thinking has usually been defined as a chain of argument, explanation, logical induction, or deduction. Engineering thinking is believed to occur by way of proposed hypotheses, suggested evidence, and rational conclusions. Other mental processes—those of meditation, intuition, belief—are usually not granted the honorific name of thinking. Yet it is evident that many complex, and sometimes profound, operations of the mind must precede our final arranging of an argument, finding a path of explanation, or staging a deduction. We have names for some of these independent operations:

- Classification,
- Reconciling, and
- Sequencing.

They are considered the underpinnings of thought rather than thinking itself, which is conceived of as an intentional set of rational linkages leading to a convincing result.

Compared with engineering thinking, poetic thinking has often been considered of an irrational nature, more expressive than logical, more given to meditation than to coherent or defensible argument. The "proofs" it presents are, it is judged, more fanciful than true, and the experiences it affords are emotional and idiosyncratic rather than dispassionate and universal. When using poetic thinking, you can express emotions and feelings and share fears, likes, dislikes, loves, and hates, which are intuitive or instinctive gut reactions or statements of emotional feeling (but not any justification). This can be applied as a tool for group discussion and individual thinking in industrial design engineering.

The additional fact that poetry is directed by an aesthetic imperative, rather than by a forensic or expository one, casts its significance on industrial design engineering.

"Play is the highest form of research," said Albert Einstein. It is also the highest form of engineering. A poem can be more lighthearted than the usual "thinking" process; it can be satiric or frivolous. High seriousness may attend it—or may not. Bizarre imaginative fantasies may be what a poem has to offer, or "nonsense," or some reduction of language that would normally be considered inadequate to "adult" thinking. Unlike the structure of a perspicuous argument, the structure of a poem may be anything but transparent, at least at first glance. The transparency facilitates simplification, the core of Lean manufacturing.

7.1   Lean manufacturing: Model T's dream cruise

The Woodward Dream Cruise (WDC) is the world's largest one-day automotive event, drawing about 1.5 million people and 40,000 classic cars each year from around the globe. It is held annually the third Saturday of August on Woodward Avenue just north of Detroit in Michigan.

There are Model T's, Model A's, muscle cars from the 1960s and 1970s, home-builds, hotrods, rat-rods, motorcycles, and so on.

A few years ago, an impeccably maintained, all original Model T was on display in Ferndale during the Woodward Dream Cruise.

Lean manufacturing is a new enterprise approach to speed, simplify, and optimize business processes in every conceivable area from manufacturing cars to treating patients.

Many of the ideas behind Lean manufacturing can be traced back 100 years to the assembly lines of Henry Ford, producing a Model T (see Figure 7.1) in only 93 minutes (cycle time).

Henry Ford's vision was to "build a car for the great multitude." The electrification of previously steam driven machinery together with new management and production techniques enabled him to take twentieth century mass production to a new level and to produce a Model T in only 93 minutes. "Any customer can have a car painted any color that he wants so long as it is black," said Henry Ford.

The problem with Ford's system was not the flow: he was able to turn the inventories of the entire company every few days. Rather, he was not able to provide variety. The Model T was not just limited to one color. It was also limited to one specification so that all Model T chassis were essentially identical up through the end of production in 1926.

The customer did have a choice of four or five body styles, a drop-on feature from outside suppliers added at the very end of the production line. Indeed, it appears that practically every machine in the Ford Motor

Figure 7.1 Horseless carriages and Ford's Model T. (Courtesy of Library of Congress.)

Company worked on a single part number, and there were essentially no changeovers.

Lean manufacturing has to be simple and pragmatic. Daft approaches like marking desks out with tape waste people's time and intellect. The fundamental requirement of Lean is to identify wasted resources.

7.2   Lean Six Sigma: Leadership is a choice

Some law other than the conduct of an argument is always governing poetic thinking, even when the poem purports to be relating the unfolding of thought. On the other hand, even when a poem seems to be a spontaneous outburst of feeling, it is being directed, as a feat of ordered language, by something one can only call thought. Yet in most accounts of the internal substance of poetry, critics continue to emphasize the imaginative or irrational or psychological or "expressive" base of poetry; it is thought to be an art, an important aspect of leading industrial design engineering. Lean Six Sigma is a methodology that relies on a collaborative team effort to improve performance by systematically removing waste, combining Lean manufacturing/Lean enterprise and Six Sigma to eliminate the eight kinds of waste (muda):

- Time
- Inventory
- Motion

- Waiting
- Overproduction
- Overprocessing
- Defects
- Skills

Many years ago, a staff member, also a Lean Six Sigma Green Belt, was going to lead a Kaizen continuous improvement event.

One week before the starting date, my staff member told me that a major stakeholder suggested that he should participate, rather than lead, this Kaizen event, since he had no leadership experience.

"Do you think that you can lead the Kaizen event?" I asked my staff member.

"If you also think that you cannot lead, then you cannot lead."

"If you think you can lead, you should tell your team and all the stakeholders."

Control factors, noise factors, and their interactions often complicate our decision-making under uncertainty. Lean Six Sigma requires both business and technical leadership skills.

Has anyone ever asked you: "What's the essence of leadership?"

Before influence, leadership is a choice you make: a choice to make a meaningful difference in the world and the lives of others. Leadership is a choice. We can choose to picture a challenge negatively or positively, which is the ultimate control factor in our mind-set. Leadership starts from our heart. At its core leadership is a choice. It's a choice to do the work of leadership. To lead requires we respond to reality, that we accept responsibility to be the change we want to see in the world, that we inspire hope in others, that we motivate others to participate in a meaningful journey of change.

Every day we're faced with the leadership choice. A choice to lead, to accept responsibility, to live authentically, to be an example to others, of the change we want to see in the world. Leadership is about our personal decision to make a difference in the world and the lives of others. It's about making a meaningful difference in the world.

All effective industrial design engineers have made a deliberate choice to lead, to do the work of leadership, to respond to our reality, to accept responsibility to be the change we want to see in the world, to inspire hope in others, and to motivate others to participate in a meaningful journey of change.

7.3 In an aircraft, changing a light bulb is an avionics engineering problem

All poems contain within themselves implicit instructions concerning how they should be read. These encoded instructions—housed in the sum of all the forms in which a poem is cast, from the smallest phonetic group to the largest philosophical set—ought to be introduced as evidence for any offered interpretation. This may help us to communicate in engineering, stimulating interests.

> Question: How many mechanical engineers are needed to change a light bulb?
> Answer: Zero; because, in an aircraft, changing a light bulb is an avionics engineering problem.

Avionics are the electronic systems used on aircraft, artificial satellites, and spacecraft. Avionic systems include communications, navigation, the display and management of multiple systems, and the hundreds of systems that are fitted to aircraft to perform individual functions. The book titled *Lean Manufacturing: Business Bottom-Line Based* presents a goal tree for developing a new light bulb. Targeting to reduce the time-to-market for a new light bulb, the goal tree includes:

- Goal 1.1: Understand the Voice of Customers (VOC) including both internal and external customers.
- Goal 1.2: Monitor and track New Product Introduction (NPI), incorporating Failure Reporting, Analysis and Corrective Action System (FRACAS).
- Goal 1.3: Improve yield of manufacturing and assembly, ensuring throughput based on Theory of Constraints (TOC)

The way thinking goes on in an industrial design engineer's mind during the process of creation and the evolution of that thinking can be deduced from the surface of the industrial design—that printed arrangement of technical ideas, forming a visible core of engineering. Similarly, a poet meditating on a given topic often thinks serially through the topic by reframing it in poem after poem, creating an active process of thinking that generates as a result the entirely different structural inner shapes those poems adopt.

7.4   On the ship of continuous improvement, "don't give up the ship ..."

On the ship of continuous improvement, "Don't give up the ship." "Don't give up the ship" remains one of the odder naval battle cries from a forgotten war, a continuation of the American War of Independence from Great Britain. It also remains one of the most inspiring phrases for continuous improvement in industrial design engineering.

James Lawrence, the captain of the Chesapeake, is said to have given a dying command of "Don't give up the ship!" Commodore Oliver Hazard Perry, a colleague and friend of Captain Lawrence, named the brig that would be his flagship, the USS Lawrence, in honor of the captain of the Chesapeake. He also had a large battle flag sewn, a blue banner with the words "DONT GIVE UP THE SHIP" stitched in white letters.

In September 1813, when Perry challenged British control of Lake Erie near Put In Bay off the coast of Ohio, British long guns did serious damage to the USS Lawrence before the Lawrence's carronades could be brought to bear. Perry hauled down his colors and was rowed through heavy fire to the brig USS Niagara where he organized his remaining schooners to press the attack against the British ships. In command of the Niagara, flying the "DONT GIVE UP THE SHIP" battle flag (see Figure 7.2), Perry captured the entire British squadron of two ships: a schooner and a sloop. This denied the British control of Lake Erie and thus cut off supplies to British forces in the field. It was one of the few significant American victories in the war. A replica of the USS Niagara is now the official tall ship of the Commonwealth of Pennsylvania. Despite having originated in the defeat of the USS Chesapeake and the loss of the USS Lawrence, "Don't give up the ship" has remained a favorite battle cry in the US Navy.

Falling down is part of growing up. Yes, on the ship of continuous improvement, "Don't give up the ship," a paraphrase of the dying words of US Navy Captain James Lawrence by US Navy Commodore Oliver Hazard Perry in the War of 1812 during the Battle of Lake Erie.

Figure 7.2 The iconic flag carried by Commodore Oliver Hazard Perry during his stirring victory at the Battle of Lake Erie on September 10, 1813. (Courtesy of the Pennsylvania Historical and Museum Commission.)

7.5 How can leaders fight short-termism for industrial design engineering?

A key performance indicator (KPI) is a business metric used to evaluate factors that are crucial to the success of an organization. KPIs differ per organization; business KPIs may be net revenue or a customer loyalty metric. However, KPIs can be risky for long-term goals and continuous improvement since they may be biased toward short-termism and the silo improvement approach.

Like recommended, the throughput accounting (TA), a key element of Goldratt's TOC, could help develop a balanced business scorecard (BBC). Here, TA is a principle-based and simplified management accounting approach that provides managers with decision support information for enterprise profitability improvement. TA is used to support a very specific managerial view of an operation—the implementation of TOC in an organization. The focal point of TA is on the incremental value of more effective employment of a constraint. TA's benefits include the following:

- Avoids local optimization. The system will not succeed if one or more individual processes are continually being rewarded for optimizing individual production. However, TA helps avoid such sub-optimization by encouraging cooperation to accomplish the organization's profit goals.
- Improves communication between departments. The focus of everyone must be on the large-scale needs of the organization. This is not possible if communication is hindered or non-existent. TA uses the Drum-Buffer-Rope process as a communication vehicle throughout the organization.

TA is a product management system that aims to maximize throughput, and therefore cash generation from sales, rather than profit. A just in time (JIT) environment is operated, with buffer inventory kept only when there is a bottleneck resource. TA for JIT is based on three concepts.

Concept 1

In the short run, most costs in the factory (with the exception of materials costs) are fixed (the opposite of ABC, which assumes that all costs are variable). These fixed costs include direct labor. It is useful to group all these costs together and call them total factory costs (TFC).

Table 7.1 Measures and consequences for improving a throughput accounting ratio

Measures	Consequences
• Increase sales price per unit • Reduce material cost per unit, e.g., change materials and/or suppliers • Reduce operating expenses	• Demand for the product may fall • Quality may fall and bulk discounts may be lost • Quality may fall and/or errors increase

Concept 2

In a JIT environment, all inventory is a "bad thing" and the ideal inventory level is zero. Products should not be made unless a customer has ordered them. When goods are made, the factory effectively operates at the rate of the slowest process, and there will be unavoidable idle capacity in other operations.

Work in progress should be valued at material cost only until the output is eventually sold, so that no value will be added and no profit earned until the sale takes place. Working on output just to add to work in progress or finished goods inventory creates no profit, and so should not be encouraged.

Concept 3

Profitability is determined by the rate at which "money comes in at the door" (i.e., sales are made) and, in a JIT environment, this depends on how quickly goods can be produced to satisfy customer orders. Since the goal of a profit-orientated organization is to make money, inventory must be sold for that goal to be achieved. The bottleneck resource slows the process of making money.

In a throughput environment, production priority must be given to the products best able to generate throughput, that is, those products that maximize throughput per unit of bottleneck resource. The TA ratio can be used to assess the relative earning capabilities of different products and hence can help with decision making. How can a business improve a throughput accounting ratio? The potential measures are shown in Table 7.1.

7.6 Statistical mean (average) can be statistically fatal

The terms mean, median, and mode are used to describe the central tendency of a large data set. Range provides context for the mean, median, and mode. When working with a large data set, it can be useful to represent the entire data set with a single value that describes the "middle" or "average" value of the entire set. In statistics, that single value is called the central tendency and mean, median, and mode are all ways to describe it. To find the mean, add up the values in the data set and then divide by the number of values that you added. To find the median, list the values of the data set in numerical order and identify which value appears in the middle of the list. To find the mode, identify which value in the data set occurs most often. Range, which is the difference between the largest and smallest value in the data set, describes how well the central tendency represents the data. If the range is large, the central tendency is not as representative of the data as it would be if the range was small. Industrial design engineers need to understand the definition of mean, median, mode, and range to plan capacity and balance load, manage systems, perform maintenance, and troubleshoot issues. These various tasks dictate how industrial design engineers calculate mean, median, mode, or range, or often some combination, to show a statistically significant quantity, trend, or deviation from the norm. Finding the mean, median, mode, and range is only the start. The engineer then needs to apply this information to investigate root causes of a problem and accurately forecast future needs or set acceptable working parameters for industrial designs.

Milton Friedman (July 31, 1912—November 16, 2006) was an American economist, statistician, and author. He was a recipient of the Nobel Memorial Prize in economic sciences, and is known for his research on consumption analysis, monetary history and theory, and the complexity of stabilization policy. It is a testament to Milton Friedman's influence and legacy that many contemporary politicians, economists, and academicians still ask, "What would Milton say?"

The mean is what people think of as "the average." To calculate mean, add together all of the numbers in a set and then divide the sum by the total count of numbers. For example, in a data center rack, five servers consume 100 watts, 98 watts, 105 watts, 90 watts, and 102 watts of power, respectively. The mean power use of that rack is calculated as (100 + 98 + 105 + 90 + 102 W)/5 servers = a calculated mean of 99 W per server. Intelligent power distribution units report the mean power utilization of the rack to systems management software. Dr. Friedman said "Never try to walk across a river just because it has an average depth of 4 feet." Remember, statistical mean (average) can be statistically fatal.

The book titled *Lean Manufacturing: Business Bottom-Line Based* presented how to develop a business management plan to mitigate exogenous shocks' influence on the business bottom-line.

7.7   Industrial engineering design—it all starts with a dream

7.7.1   Kindergarten classrooms: Where engineering dreams start

Poetic thinking is about finding inspiration from different facets of life in aid of the production of an industrial engineering design, as well as helping the functional design achieve an emotional impact as an engineer's best inspiration. Let's start from the story about the Mother's Best Flower. Lisa always expects flowers from our kids, especially on Mother's Day. However, I believe the best flowers for any mother is her kids live an impactful life that inspires others to live with more passion, more love, and renewed vision.

My inspiration starts from my son Sonny's story, starting from last year's Mother's Day. My wife Lisa checked the door of our home many times, hoping the flowers from Sonny would magically appear by the door. Unfortunately, both Lisa and I knew we could not receive flowers, because Sonny died tragically in a fire accident one and half months before. However, it was also on that day we received the good news of his church missionary. That good news reminded me of a childhood story about the best flowers for a mom.

Many years ago, a young man was going away for his mission, dangerous and noble. Before leaving his mom, he said, "Mom, I will send my horse with the best flowers to you if I cannot return to you, to my sweet home."

The mom sealed his mouth with her hand, and said, "No, son. If I can hear the good news of your accomplished mission, if the horse can carry the good news back to our sweet home, that is the best flowers for me."

Magically, we have been always receiving good news of Sonny's Mission on Mother's Day:

- Three years ago, we received the news that Sonny accomplished his trip to Wisconsin as a Church's youth director.
- Two years ago, we received the news that Sonny accomplished his trip to Nebraska to help the Native American community there.
- Last year, after Sonny passed away, we received the news about Sonny Wang's Kindergarten from the Mission Starfish, a church in Haiti.

It was on Mother's Day last year, on the bank of the beautiful Grand River, Lisa and I read a letter about Mission Starfish. With the help of money from Sonny's life insurance policy the Mission was able to build the Sonny Wang Kindergarten, established in Haiti to honor our Sonny who extended his love far beyond his church and community. As shown by Figure 7.3, in Haiti, the beautiful Sonny Wang Kindergarten has classroom space for 135 kids and worship space for over 100 people!

According to research, engineering dreams start from kindergarten classrooms. Concepts about the world—the beginning of engineering—begin at birth. Young children, particularly kindergarten-aged children, have inquiring minds and are natural engineers. Kids enter kindergarten classrooms with curiosity and the ability to explore. These make them enthusiastic about learning about our world. They wonder about

- How things work,
- Why things change, and want to experiment, touch, and
- What happens if ….

As shown by research at the University of Maryland (see Figure 7.4), where Sonny attended the Center for Young Children (CYC) many years ago, learning about engineering builds on this period in kindergarten children's development. Engineering offers children the opportunity to do what comes naturally:

- Observe,
- Ask questions (what, how, and 5 whys …),
- Manipulate objects,
- Communicate their thinking through actions, words, drawings, or constructions, and
- Build things together (group technology).

Figure 7.3 Sonny Wang's kindergarten. (Courtesy of Mission Starfish Haiti.)

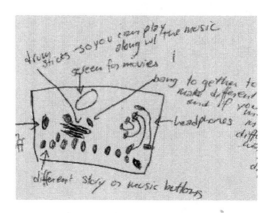

Figure 7.4 Sketch by two 5-year-old children, showing what the children would like the "MusicBlocks of the Future" to be like. (Courtesy of https://www.cs.umd .edu/hcil/kiddesign/cof.shtml.)

It's in kindergarten classrooms that kids get the early ideas: engineering is a way of doing. Engineering is solving problems, using a variety of materials, designing and creating, and building things that work.

On the bank of the river, I recalled it was here that Sonny and I talked about:

- Never too old to learn, and
- Never too young to learn.

Sonny, were you talking about future kindergarten classrooms, where many engineering dreams would start?

7.7.2   The starfish and continuous improvement: Every action, no matter how small, can make a difference

Today the Sonny Wang kindergarten building comes to our world as Creole-based digital tools are entering classrooms to improve science, technology, engineering, and mathematics (STEM) education in Haiti. Historically, Haitian children have been educated exclusively in French, a language in which most of the population are not fluent. Using Creole for Haitian education will provide Haitian children quality access for STEM education.

While Haitian children feel most comfortable with Creole, the use of French in Haiti's classrooms has been a national education policy. School exams as well as national assessment tests are mostly conducted in French, rather than Creole. STEM course materials have been available exclusively in French, too. In Haiti's classrooms, most children do not like to ask or answer questions because they are constantly struggling to translate from Creole into French or from French into Creole.

The use of French creates problems for teachers as well. Haiti's teachers prefer to teach in Creole because that is the language with which they feel most comfortable also. They like to make jokes when they teach. That humor is essential for good teaching—to wake the students up, to keep them alert, and to make them feel relaxed.

Now the work of pro-Creole educators both in Haiti and in the Haitian Diaspora starts to show the key benefits of a Creole-based education at all levels of the education system. Earlier this year, Haiti adopted a new educational policy that will allow students to be educated in Creole, which is as capable of conveying complicated intellectual concepts as any other Indo-European tongue.

Creole-based digital tools meet crucial needs in Haiti by introducing modern techniques for interactive pedagogy while helping to develop digital resources in Creole. Digital tools including STAR, Mathlets, and PhET have been translated into Creole and provide proof of concept of Creole as a necessary ingredient for active learning in Haiti.

The initiative of using Creole-based digital tools will have a profound impact on the way people think about teaching STEM in mother tongues, and serve as a very important model for similar initiatives around the globe. Across large swaths of Africa and the Americas, indigenous languages continue to face systematic marginalization. This new initiative provides a guide for these populations to empower their children with engineering tools to mitigate risk and uncertainty in STEM education.

On the bank of the beautiful Grand River, Lisa and I read about Sonny Wang's kindergarten classrooms, where beautiful engineering dreams will start. Looking at dreaming a creative tool—a catalyst—for productivity and problem solving, the new kindergarten will show explored free flow of thoughts as a design method, how daring to dream leads to final creative output.

For industrial design engineering, continual improvement process is an ongoing effort to improve products, services, or processes. Sonny lived an impactful life with a renewed vision reflected by the Mission Starfish:

- Every action, no matter how small, can make a difference.
Here is the Mission Starfish Story as told to me:

> A young boy walked along the beach and found thousands of starfish washed up because of a terrible storm. When he came to each starfish, he would pick it up and throw it back into the ocean. People

watched him with amusement, and said, "Boy, why are you doing this? You can't save all these starfish. There are too many. You can't really make a differ-ence!" But the boy continued to bend down, picked up another starfish, and threw it as far as possible into the ocean, and said, "Well, I made a difference to that starfish."

By conclusion, let me call to action—for each of you, you never know what day will be your last, so live an impactful life that will send the good news of your mission to your mom. That is the best flowers you can send to your mom just as Sonny sent the best flower to his mom Lisa.

7.8   *Manufacturing excellence flows with robust design, just like a great river*

Speaking at Sonny's memorial service, I remembered, for celebrating his twenty-fifth birthday on March 19 this year, his journey back to the beau-tiful Grand River by which our sweet home is located.

I also recalled our journey together when I was writing the book titled *Lean Manufacturing: Business Bottom-Line Based*. When Sonny was driv-ing over interstate I-74 covered by heavy snow, I wrote down the book's Preface:

> About 50 miles east of the Mississippi River, I found myself back in Galesburg, a western city of 34,000 and the birthplace of poet Carl Sandburg.
>
> ...
>
> I found Carl Sandburg College's Center for Manufacturing Excellence, representing the beauti-ful city's continued pursuit for world-class manu-facturing.

Manufacturing excellence flows with robust design, just like the great Grand River (Figure 7.5) and Mississippi River.

Bridges are subject to a variety of factors during their lifetimes, which could lead to structural failure. Those factors include boat impacts, earth-quakes, severe wind loads, floods, material defects, and flaws in design. Depending on the bridge design and location, each of these factors has a probability of occurrence. Based on these probabilities, it may be possible to predict the likelihood of the bridge's collapse.

There are several reports regarding the analysis of collapsed bridges and their causes.

Figure 7.5 Robust truss design: My impression of the 6th Street historic bridge over the Grand River (Grand Rapids, Michigan).

For instance, the report on the 1-35W steel truss bridge in Minneapolis, Minnesota, clarifies that the dead load of the structure had increased due to repair and slab reinforcement. Additionally, the thickness of the gusset plate that failed was half of its design value. These factors contributed to the bridge's collapse. Similar investigation studies have been conducted on existing structures, particularly bridges that are aging or deteriorating.

The disastrous structural collapses that have resulted in the past from local failures of critical members have emphasized the significance of designing structures to be resilient. This notion of designing structures to withstand extraordinary loads was first brought about with the partial collapse of London's Ronan Point Apartment Tower in 1968. The structure's lack of alternative load paths prevented the redistribution of forces after an internal explosion occurred.

Designing structures to be resilient to extreme loads has become a topic of interest in recent years, which has been triggered by the progressive collapse of structures in the past. Structural failure due to the lack of resilient design has been particularly prevalent in bridges. The failures have been results of a variety of factors that the bridges have been subjected to. The objective of preventing the occurrence of future collapses has encouraged further research into the design of resilient structures. To prevent future occurrences of collapse, engineers have been researching methods to design structures to be resilient. Two methods of achieving resilience are by incorporating structural redundancy and robustness in bridge design.

7.8.1   Robustness

Robustness is defined as the ability of a structure to continue to carry load after being brought to a damaged state. This means the structure will continue to stand after an event damages a part of the structure. Essentially, the damage that is caused to one part of the structure is not propagated to the rest of the structure. This property of a "robust" structure, when coupled with a continuous or periodic inspection program using nondestructive evaluation (NDE) techniques, is useful in failure prevention, because such a structure is expected to display "measurable" signs of "weakening" long before the onset of catastrophic failure.

7.8.2   Redundancy

Redundancy is defined as the ability of a structure to continue to carry load after the failure of a single element. The forces that could no longer be taken by the removed element would then be redistributed to the surrounding structure. Redundant designs generally incorporate structural members that are not strictly necessary for the design loads, but are essential in an extreme event.

Robust bridge designs are generally more effective for bridges with longer spans, whereas designs with redundancy are better suited for shorter spans. As the amount of structural damage that is incurred by a bridge increases, the more redundancy should be built into the structure.

7.9   We have the freedom to define our success

7.9.1   Personal branding and success in industrial design engineering

During an international flight, Sonny asked me: "Dad, are you successful?"

As a risk engineering professional having defined "success criteria" for many large-scale systems, I answered: "Who would define our success? We have the freedom to define our success."

When Lisa, Cheney, and I celebrated Sonny's life with hundreds people, I told Sonny: *"You are successful because you have done so many good things with all your love and all your life, which we are truly proud of."*

Personal branding is important for success in industrial design engineering, which requires the courage to invent, to innovate, and to improve. Each of us is also blessed with the freedom to define success. Success is different for each of us, and fearless brands have realized the freedom to define success as they determine—free from the influence of others' expectations. As illustrated in Figure 7.6, this gives us the freedom

Figure 7.6 We have the freedom to define our success.

to define success on our terms. And, we can redefine success at any time to reflect our own growth.

A truly fearless brand has achieved a level of true humility. Humility in this case means accepting our faults—and our strengths—for what they are—without exaggeration or minimization. This clarity leads to the conviction of self—which in turn creates tremendous freedom.

7.9.2   Fearless brands enjoy 7 degrees of freedom

Building a fearless personal brand allows each of us to create freedom. That freedom manifests in a variety of ways and opportunities.

1. Freedom to stand out: One of the most obvious reasons to build and embrace your personal brand is to stand out. With over 7 billion people in the world, it is essential to stand out. As Stanley Marcus (Neiman-Marcus) says, "The dollar bills the customer gets from tellers in four banks are the same. What are different are the tellers." To succeed one needs to stand out.

2. Freedom to fit in: An often overlooked aspect of branding is realizing how your personal brand fits in—be it at a company, a group, a family, or in public. A fearless brand has enough humility (see above) to know and accept its role.

3. Freedom to define success: Success is different for each of us, and fearless brands have realized the freedom to define success as they determine—free from the influence of others' expectations.

4. Freedom from judgment: There is no shortage of people, advertisements, articles, movies, and such which are quick to state what you

should be, how you should dress or act, what you should be doing. A fearless brand defines itself.

5. Freedom from the fear of no: When one knows one's purpose and relevance, one is able to accept the word NO. You realize that your value proposition doesn't satisfy a particular situation and is not a reflection on your validity.

6. Freedom from distraction: Having clarity of purpose has an oft over-looked benefit—knowing what does not move you closer to your goals. That understanding lessens the likelihood of pursuing what seems like great opportunities but in reality are mere distractions.

7. Freedom to be: Yes, the freedom to be your truest self—to just be. Is there any greater freedom?

Fearless brands determine success by their own standards. Likewise, they create a great deal of freedom by determining their why, embracing their passion, understanding their talent, and by knowing what is—and what is not—relevant. Think of the 7 items above as the fearless branding 7 degrees of freedom.

7.10   *Probabilistic industrial design: Probably we can have a statistical Black Friday*

In this section, we introduce the field of probabilistic robotics and its link to Black Friday, the day following Thanksgiving Day in the United States (the fourth Thursday of November), often regarded as the beginning of the Christmas shopping season.

During the holiday season, robot art is an important part of a Black Friday feast. Robotic control is an exciting and highly challenging research focus. The information acquired using robot sensors is noisy, incomplete, and ambiguous. Currently, most robotic control methods can only adopt a PC-based microcomputer or the digital signal processor (DSP) chip to real-ize the software part or adopt the field programmable gate array (FPGA) chip to implement the hardware part of the robotic control system. They do not provide an overall hardware/software solution by a single chip in implementing the motion control architecture of the robot system.

An FPGA is an integrated circuit designed to be configured by a cus-tomer or a designer after manufacturing—hence "field-programmable." For the progress of VLSI technology, the FPGAs have been widely inves-tigated due to their programmable hard-wired feature, fast time-to-market, shorter design cycle, embedding processor, low power consumption, and higher density for implementing digital systems. FPGA provides a tradeoff between the special-purpose application specified integrated cir-cuit (ASIC) hardware and general-purpose processors. The novel FPGA technology is able to combine an embedded processor intellectual property (IP) and an application IP to be a system on-a-programmable-chip (SoPC) developing environment, allowing a user to design an SoPC module. The circuits required with fast processing but simple computation are suitable to be implemented by hardware in FPGA, and the highly complicated control algorithm, reducing development time and cost.

Probabilistic robotics, also called statistical robotics, is a field of robotics that involves the control and behavior of robots in environments subject to unforeseeable events. Because reality always involves uncertainty, probabilistic robotics may help robots to more effectively contend with real-world scenarios. Originally, probabilistic robotics involved the ability of a robot to locate itself using maps of known work environments. A blueprint of the surroundings, along with tools such as proximity sensing and machine vision, allowed a robot to navigate and perform tasks with a minimum of errors or mishaps. More recently, probabilistic robotics has become concerned with the development of robots that can work effectively in environments that they have not previously encountered. Therefore, a robot must develop a sense of the most likely result of a given movement or actions, based on defined statistical functions, and then strive for the optimum outcome.

In summary, robots create models of environment using probabilities and solve tasks with the help of probabilistic reasoning. The parameters of probabilistic models can be learned using the environmental data.

Probably someday we will have a statistical Black Friday feast.

7.11   Industrial engineering thinking: Experimenting with plot to characterize life's unfolding

Engineering is an extremely exciting profession. Engineers are behind many of the advancements we take for granted today including computers, the Internet, smart phones, medical tools, and cars. Engineers use science and math to create cool products and machines that make life an enjoyable living.

Engineering is not only an extremely exciting profession. It is a way of thinking. The word "engineer" is from the Latin word *ingenium*, which shares the same root as the word "genius." Certainly, every engineer would like to find optimal solutions to practical problems. However, the constraints of the problem-solving are of several, qualitatively different types. Often there is no formal way to find the best trade-offs.

7.11.1   Challenge to engineering optimization: NP hardness

For example, computational complexity presents a fundamental challenge to engineering optimization due to NP hardness. A problem is NP-hard if an algorithm for solving it can be translated into one for solving any NP-problem (nondeterministic polynomial time) problem. NP-hard therefore means "at least as hard as any NP-problem," although it might, in fact, be harder.

7.11.2   *Chip: Computing exact wire length is an NP-hard problem*

For green electronics manufacturing, a chip is composed of basic elements, called cells, circuits, boxes, or modules. They usually have a rectangular shape, contain several transistors and internal connections, and have at least two pins in addition to power supply. The pins have to be connected to certain pins of other cells by wires according to the netlist. A net is simply a set of pins that have to be connected, and the netlist is the set of all nets.

The basic placement task is to place the cells without overlaps in the chip area. A feasible placement determines an instance of the routing problem, which consists of implementing all nets by wires. The quality of a placement depends on the quality of a wiring that can be achieved for this instance. Creating environmentally sensible products, it is usually good if the wire length, the total length of the wires connecting the pins of a net, is as short as possible. The power consumption of a chip grows with the length of the interconnect wires, as higher electrical capacitances have to be charged and discharged, dissipating heat to the environment. From a risk engineering perspective, signal delays increase with the wire length, impacting system integrity. Critical nets should be kept particularly short.

Thus, an important question is how to compute wire length without actually routing the chip. First, note that nets are wired in different layers with alternating orthogonal preference direction. Therefore the 1-metric is the right metric for wire length. An exact wire length computation would require finding disjoint sets of wires for all nets, which is an NP-hard problem (see Figure 7.7).

7.11.3   *Seek best solutions under constraints*

To optimize engineering problem-solving, engineers are thinking similar to mathematicians in terms of problem formation. However, in many branches of mathematics, finding the best solution is not essential. For a mathematician, finding any proof is often a success. Engineers, by contrast, are not satisfied with existence proofs. Getting something to work is inadequate. The product has to best satisfy customer requirements. Even in simple problem solving, the engineer looks for evidence that the space of possible solutions was adequately searched, and the chosen solution validated to be optimal.

Facing the challenge of NP-hardness, engineers provide explanations to justify their choices based on their judgments. Engineering thinking

Figure 7.7 Chip: Computing exact wire length is an NP-hard problem. (Courtesy of National Aeronautics and Space Administration http://www.nasa.gov/ima ges/content/194232main_MultiSpacewireConcentratorCard%201.jpg.)

develops explanations that identify and validate the problem solving under constraints. Engineering thinking involves induction (analogical reasoning) as well as deduction. Compared with science, engineering has often been considered an irrational genre, more expressive than logical, more meditative than given to coherent argument. All innovative engineers are thinkers. While the styles of innovative engineering thinking are distinctive, they are characterized by:

- Modeling: model-based ideas.
- Rethinking: reflection-based innovation.
- Experimenting: from rearranging to renovation.
- Picture thinking: from image to assertion.

In engineering decisions concerning constraints and uncertainty, the engineer draws on similar previous problem-solving. Analogical reasoning is actually at the heart of engineering thinking, requiring a rich set of source analogs from which to reason, similar to how a poet experiments with plot to characterize life's unfolding.

7.12   Quantum security of smart-card

Power analysis is a successful cryptanalytic technique which extracts secret information from smart-cards by analyzing the power consumed during the execution of their internal programs. The attack is particularly dangerous in financial application in which users insert their smart-cards into teller machines that are owned and operated by potentially dishonest entities.

In this section we describe a new solution to the problem, which is based on an application of quantum cryptography with quantum entanglement for smart-cards.

An interesting application of quantum cryptography for smart-cards using quantum transmission of photons and an application of quantum key distribution for the quantum identification system was published. Here we present a new application of quantum entanglement for smart-cards to eliminate the new physical attacks on the smart-card.

The necessity to look for more secure smart-cards follows as the consequence of the fault case presented in the *New York Times* headline or new attacks via power and differential power analysis including simple power analysis, where the eavesdropper tries to recover information about the secret key by simply measuring the power consumption of the computing device, and the more complex differential power analysis additionally requires knowledge of cipher-text outputs and is thus more costly.

Generally we can summarize that there are the following basic problems with existing identification systems using smart-cards:

- The customer must type the personal identification number (PIN) to an unknown teller machine which can be modified to memorize the PIN;
- The customer must submit the smart-card with information needed for the identification to an unknown teller machine; in the presence of an eavesdropper it can also be memorized together with the PIN.
- The smart-card based on silicon technology can be attacked even without its interaction with the teller machine via many noninvasive physical attacks.

In this way, such identification system via smart-card can fail. The solution of these problems is to employ optical fibers and optoelectronics on the smart-card together with entangled quantum optical states.

Here we present a new identification system, which in principle can be based on quantum entanglement of two photons and which solves these basic problems in the following way:

The PIN will be typed directly to the card for activation of the optoelectronics devices located on the smart-card and no PIN information will be exchanged with the teller machine.

The information needed for the identification of the card inside the teller machine will be protected against an eavesdropper through non-local projection of the state of one member of the pair of entangled photons which will happen during its propagation directly on the smart card.

The power source is put directly on the smart-card (e.g., a photocell). It is well known that in the ordinary teller machine without quantum channels all carriers of information are physical objects open to copying or cloning. The information for the identification of the card is enclosed by modification of their physical properties. The laws of the classical theory allow the dishonest eavesdropper to measure and copy the cryptographic key information precisely.

This is not the case when the nature of the channels is such that quantum theory is needed for their description. Any totally passive or active eavesdropper operating on the quantum channel can be detected.

The current development of technologies of fiber and integrated optics makes it possible to construct quantum optoelectronic smart-cards for our application. The travelling distance for quantum transmission is very short here. This means that the problems appearing in application of ordinary quantum cryptography in optical communications are negligible here.

The new idea presented here is that the projection on a selected polarization basis of the entangled photon from the Einstein-Podolsky-Rosen (EPR) pair of polarized photons occurs directly inside the smart-card, at the moment when the second member of the pair is measured in the same basis inside the teller machine.

7.12.1 The quantum entanglement for the smart-card

A scheme of the QC smart-card with the quantum entanglement is given in Figure 7.8. Here the solid line presents the quantum channel (optical

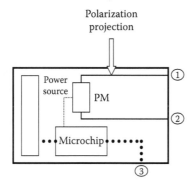

Figure 7.8 Schematic representation of the QC smart-card.

fiber), the double dotted line presents the electric connection from the photocell source and signal connection with the teller machine. The single dotted line is the signal pulse connection between the microchip and polarization modulator, which are integrated and with active shielding are "secure" against external measurements of electromagnetic fields from signal pulses. An arrow indicates the position of the correlated photon on the smart-card at the instant of the measurement of the other member of the pair in the teller machine.

This quantum card consists of:

- A PIN activator, which can have the form of the ordinary card light-source calculator with sensor keys; the user puts the PIN on the card to identify himself in a secure distance from the teller (i.e., unknown quantum cryptographic verification device) and in a secure outer area; the card will be blocked when the sequence of the three incorrect PINs is given; as soon as the activated card is not used within a short time period (say, 20 s), the activation is closed;
- A microchip with the implemented cryptographic key $\{0,1\}n$ which is long enough to yield required security for the given protocol; the microchip is activated when the correct PIN is entered on the card;
- A polarization modulator (PM), which transforms the cryptographic key $\{0,1\}n$ to the optical polarization states of the photon under the following encoding rules:
  - "0" is encoded as no change of the polarization;
  - "1" is encoded as change of the polarization state to the other basis state.
- A quantum channel consisting of a single-mode optical fiber.

The smart-card is equipped with the optical connectors for connecting the quantum channel to the teller machine and the electric connector that serves for auxiliary communication needed to trigger and synchronize the operation of the smart-card with the arrival of photons from the teller and possibly also for the exchange of classical information needed for realization of the cryptographic protocol.

Present technologies give the possibility to construct the card without significant radiation of electromagnetic energy and a new generation of microchips, which together with optoelectronics elements are resistant against possible known physical attacks. The main advantage is that no classical secret information ever leaves the smart-card; only quantum information is going out.

In this way the smart-card controls the polarization of the photon which was projected randomly at one of the two possible polarization states directly on the smart-card and no eavesdropper inside the card slot has a chance to read-out the authentication information without being detected.

7.12.2   The teller machine with quantum entanglement

The teller machine, which is based on the quantum correlation principle, is plotted in Figure 7.9 and consists of:

- Source of EPR photons; a pulsed laser source can be used to pump the down-conversion process in a nonlinear crystal;
- The delay line that determines the moment of the projection of the polarization state of the correlated photon on the smart-card at the moment of the detection of the second member of the pair in the teller machine;
- The polarization controller serves for compensation of the polarization changes in the device;
- Wollaston prism I serves for random projection of the polarization of one of the photons while Wollaston prism II is used to analyze the resulting polarization state of the other photon;
- The two pairs of photon-counting detectors are used to detect the outputs of the Wollaston prisms. The first pair of detectors (A, B) detects the realization of random projection in the Wollaston prism

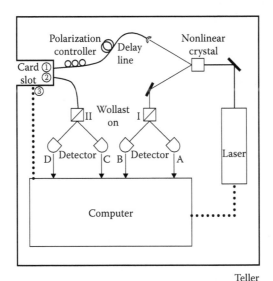

Figure 7.9 Scheme of the quantum teller machine based on the quantum correlation.

I for the photon that remains inside the teller machine. The second pair of detectors (C, D) is used to measure the state of the second EPR photon which is coming modified or not-modified from the smart-card. Via modification of the polarization smart-card sends its secret;

- The computer that drives the operation of the whole device and where the information obtained by measurement on the detectors A, B, C, D is processed according to the given cryptographic protocol and the authentication key is compared to its stored replica;
- All quantum optical paths inside the teller machine can be done with ordinary optical single-mode fibers (solid line) and electric signal channel (dotted line) are in the same quality like on the smart-card.

In such realization, the smart-card controls polarization while the teller machine controls detectors and can obtain the secret information from smart-card only via quantum transmission. The teller machine accepts or rejects the smart-card if the secret information coincides with the replica stored in its database up to a tolerable amount of errors. Some amount of errors must be tolerated due to physical imperfections of the device.

This is the authentication of the smart-card, which can be improved via known cryptographic protocols that combine sending the classical public information, which can be under eavesdropper's control, and quantum secret information, where eavesdropper's activity will be detected.

We have discussed the possibility to utilize the advantages of quantum correlation for authentication. Of course, if we have a more robust smart-card with detectors we can provide the mutual identification between smart-card and teller machine via quantum cryptography.

Manufacturers should be encouraged to develop enough cheap photon-counting detectors and smaller optoelectronics elements integrating the new generation of microchips with polarization modulators.

We have shown a new positive solution to the open problem connected with the smart-card security.

chapter eight

Risk communications and continuous improvement
Poetic process engineering

> The spiritual mind is always metaphorical. Spiritual thinking is poetic thinking. It's always trying to put a very diaphanous experience into words, realizing all the while that words are inadequate.
>
> **Sam Keen**

8.1 Business communication as a critical element of industrial design

As technologies evolve, technical writings are not just presented in 2D documentation with text and drawings of individual discipline. Many times, it is so hard to explain complex products in 2D linear drawings. Since the world becomes flat, engineering professionals have to think far beyond the traditional boundaries of technology, incorporate nonlinear lateral thinking and 3D visual presentation to satisfy consumers' dynamic needs globally, balancing short-term markets and long-term trends proactively.

8.1.1 Rhetorical question and business communication

"Auld Lang Syne" has influenced us for generations. The first page of the Manuscript Division's photographic reproduction of the "Auld Lang Syne" manuscript is shown in Figure 8.1. At the end of *It's A Wonderful Life*, George Bailey realizes he wants to live. And everyone from the mythical town of Bedford Falls comes barreling into his living room to tell him how much they love him and sing "Auld Lang Syne."

In the first *Sex in the City* movie, "Auld Lang Syne" plays while Carrie rushes to meet Miranda before midnight hits on New Year's Eve.

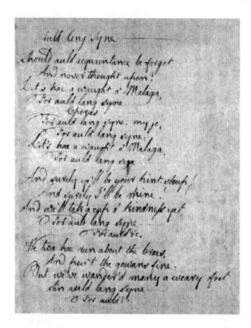

Figure 8.1 The first page of the Manuscript Division's photographic reproduction of the "Auld Lang Syne" manuscript. (Courtesy of Library of Congress.)

8.1.2 "Should old acquaintance be forgot?" is a rhetorical question

Why is "Auld Lang Syne" so influential? A rhetorical question plays an important role:

- "Should old acquaintance be forgot?"

As immortalized in *When Harry Met Sally*, a casual listener to the song "Auld Lang Syne" is likely to be confused as to what the central opening lyric means:

Harry: [about Auld Lang Syne] What does this song mean? My whole life, I don't know what this song means. I mean, "Should old acquaintance be forgot"? Does that mean that we should forget old acquaintances, or does it mean if we happened to forget them, we should remember them, which is not possible because we already forgot?
Sally: Well, maybe it just means that we should remember that we forgot them or something. Anyway, it's about old friends.

The answer is that it's a rhetorical question. The speaker is asking whether old friends should be forgotten, as a way of stating that obviously one should not forget one's old friends.

A rhetorical question is a figure of speech in the form of a question that is asked in order to make a point rather than to elicit an answer. It's a question asked solely to produce an effect or to make an assertion and not to elicit a reply. A rhetorical question is asked just for effect or to lay emphasis on some point discussed when no real answer is expected.

A rhetorical question may have an obvious answer but the questioner asks rhetorical questions to lay emphasis to the point. In literature, a rhetorical question is self-evident and used for style as an impressive persuasive device. Percy Bysshe Shelley ends his masterpiece "Ode to the West Wind" with a rhetorical question.

O Wind,
If Winter comes, can Spring be far behind?

The poet achieves the desired effect by asking this rhetorical question instead of making a statement. The answer to this question is not sought; rather, an effect is successfully created giving a fine finishing touch to the ode.

Broadly speaking, a rhetorical question is asked when the questioner himself knows the answer already or an answer is not actually demanded. So, an answer is not expected from the audience. Such a question is used to emphasize a point or draw the audience's attention.

Rhetorical questions are questions that are made for a purpose but not supposed to be answered. When you're asking a rhetorical question that is more of an obvious answer, it causes you to think about what was said, even pausing for the start of a discussion. Here is a quote from Fred Friendly:

> Our agenda... is not to make up people's minds but to open minds to make the agony of decision-making so intense that the only escape is thinking.

Thinking rhetorically you will always have a purpose for your business communication. You will know exactly what you are trying to express. In order to be good at business communication, we must truly understand the meaning of the word rhetoric and apply it in our everyday life. By thinking rhetorically you could say you're acting like an engineer. Paying close attention to every detail on design specifications, listening closely to your customer's requirements, and then once you have all the information then thinking, designing, and developing critically about the things you have learned. You are also trying to persuade the design review team to believe what you believe. In summary, thinking "rhetorically" means reasoning with audience predispositions in mind.

8.2   Rolling into 2016: Brevity is the soul of business communications

Having been stuck inside of my SUV on the road to Memphis Rock 'n Soul Museum for hours, I was wondering how a good rock song can move a generation to action. While rock 'n roll music entered the popular music spectrum in the 1950s, rock music really came into its own in the 1960s. In the 1960s, rock music came of age and dominated the popular music charts. As the genre grew and changed, many diverse and new subgenres emerged, all tied to original rock but each with their own unique style and purpose. At that time, rock was a voice of unity and liberty saturated not only with ideals of defiance but also with dreams of love, community, and spirituality. In the 1960s songs became the way for a generation to communicate its angst and hopes to each other and to the world. The result was a decade of change and transformation that defined an era. The Beatles, Bob Dylan, Peter, Paul and Mary, and others wrote and sang the words that inspired social action on a hitherto unknown scale.

They did so with 3- to 4-minute songs that incorporated and encapsulated strong and direct messages—messages that struck their audiences with a force that belies belief. And they accomplished a global social revolution without the ubiquitous social media we have become used to using in today's global village to get our messages out. In 1962, Bob Dylan released his first important original song, "Blowin' in the Wind," which Dylan claims that he wrote in about 10 minutes one afternoon. It was released on his first album of all original songs, *The Freewheelin' Bob Dylan*. The song was a breakthrough song for so many reasons. First, it was a catchy rock song, pure and simple. Second, it had tremendous depth lyrically, well before it was common for rock songs to have any depth at all. And third, was its stunning simplicity and brevity. Here, science and humanity linked as one eternity.

Great songs reveal their depth more and more with each listen. And though I'd heard "Blowin' in the Wind" dozens, if not hundreds, of times before, when I listened to it in my SUV on the way to Memphis, it was the first time that I noticed how uniquely and simply the song was constructed. Dylan's formula was simple—invented by songwriters and poets from time immemorial. Simple strong words strung together in short, pointed sentences, using alliteration when possible, and a beat that resembled the cadence of a conversation that made memorization as easy as pie. A complete story told in a few minutes that grabbed the soul and the imagination and stirred both to action, which is the essence of business communication.

In "Blowin' in the Wind," Bob Dylan asked a few simple questions in a way that shook the core of our beliefs and set a generation on a course that would change the world forever. "Blowin' in the Wind" is nothing more than nine questions, each with the same answer. These nine straightforward questions were about life, war, love, happiness, and philosophy. No small, inconsequential issues. Yes, they are very heavy stuff that could take books to write about. Yet Dylan seemed to capture the core essence of these massive complex topics with a perfectly written, incisive question.

> How many roads must a man walk down before
> you call him a man?
> How many seas must a white dove sail before she
> sleeps in the sand?
> Yes, and how many times must the cannon balls
> fly before they're forever banned?
> Yes, and how many years can a mountain exist
> before it is washed to the sea?
> Yes, and how many years can some people exist
> before they're allowed to be free?
> Yes, and how many times can a man turn his
> head and pretend that he just doesn't see?
> Yes, how many times must a man look up before
> he can see the sky?
> Yes, how many ears must one man have before he
> can hear people cry?
> Yes, how many deaths will it take till he knows
> that too many people have died?

Each of the nine questions has the same answer:

The answer my friend is blowin' in the wind
The answer is blowin' in the wind.

The same formula can be used today to get a message across to a target or mass audience. Simple words, short sentences, a short story, and good cadence can make any speech, business presentation, or speaking note memorable to an audience and motivate people to action. For "Blowin' in the Wind," Dylan did not waste a word. Not a sentence. Not a thought. He doesn't meander through buckets of words in order to get his point across. His point came across in each syllable. He asks nine questions, and each question makes a stunningly poetic, deep, and even obvious point about humankind. But he knew that in order to get these points across to the public, he had to be brief and efficient. He had to choose carefully. He could've asked these questions in any number of ways, but what he did was challenge himself to ask these deep, loaded questions with as few words as possible.

Because he succeeded in doing so, the song is perfect. And the answer to each of the nine questions—specifically that the answer is blowin' in the wind—is also a perfect, crisply written answer to these truly unanswerable questions. There is no better way he could've answered these questions but the way he did. Brief and to the point. He had no doubt in his mind what the answer to these questions was. And yet, the content of the answer—that the answer is blowin' in the wind—by its very definition, suggests uncertainty!

So it is because of the efficiency and brevity of the words that this rock song has such impact, and will forever be known as one of the most perfect songs ever penned.

Brevity is not merely, as the saying goes, "the soul of wit." It is also a necessity in the over-inundated world of business communication skills. Times have really changed in recent years, and that affects how we communicate and process information. How can you achieve brevity for business communication? Business communication is founded on the principles of brevity. There is little room for academic loquaciousness. This applies to not just the length of your message, but also its contents. Try to use short sentences and short words. Avoid jargon and words that send the reader to the dictionary. Adopt this principle for intra-team as well as customer-focused communication. To keep communications brief, the following five strategies are recommended:

- Don't over-explain.
- Use the Ws (why, where, who, etc.).
- Replace words with images.
- Use pauses more.
- Use a "mind map" (to plot how things will fit together) before you go in.

In summary, the argument has always been that "brevity is the soul of wit"; turns out it's the soul of business communication, too. Before you do anything around your engineering work—consider any software, write a marketing campaign, design a circuit layout, institute any process, and so on—consider brevity. This is just like Dylan got to the point in each of his nine questions in "Blowin' in the Wind," quickly and with great impact.

8.3   "Take Me Home": Four things country music teach us about business communication

Hitting the road to Nashville with my family, I recalled singing country music in my university dorm with my favorite guitar (see Figure 8.2). Whether it's at home or at work, good communication is about creating connections. When this happens, you touch a chord that is personal and real. Your audience is there with you, feeling what you feel, thinking what

My favorite guitar

Figure 8.2 "Take Me Home": Four things country music can teach us about business communication.

you're thinking. For just a moment, they have stepped into your shoes. This is when you prevail over everything else and for at least a moment or two, this connection is everything. That's what country music does by using heartfelt emotion, simplicity, and a relatable story. While many people turn up their nose at country music, especially in the urban-oriented world of online communications, I believe it has valuable lessons to teach us all. Let's break it down. Here are at least four things about business communication you can learn from country music:

8.3.1 Tell a story

If you want your audience to relate to your message, tell a story. When you're done, your audience should say: "That could be me," or "I've been there," or "I want to be that person." When the audience emotionally connects with your story and sees your humanity, your message makes more sense and sticks. Have you heard John Denver's "Take Me Home (Country Roads)." "Drivin' down the road, I get a feeling that I should have been home yesterday." A bittersweet road trip story if ever there was one. Since 1971, it's a sure bet that anyone journeying through West Virginia has had this lovely, haunting hit running through their mind while traveling those scenic country roads. The magic of this song is that the story is so vivid you experience it for yourself.

Your story can be serious or funny, but it must be relatable and emotionally honest. If you're going to use a story as a central part of your communication, you need to know it cold and practice it in front of others you trust. Ask for their feedback to see if it's having the impact and meaning you're going for. Your story can be a powerful moment for your audience and by developing it you will accelerate your growth as a communicator.

8.3.2   Hook them—establish connection

Have you ever had a song you can't get out of your mind? You want to stop thinking about it, but you can't. The hook keeps running through your brain. Country music uses great hooks. This communication tool isn't self-conscious; it doesn't try to be intellectual. It's simple and you get it right away. Maybe you've heard Rascal Flatts' "Life Is a Highway"? "Life is a highway, I wanna ride it all night long/If you're going my way, I wanna drive it all night long." It's a long, long way from Memphis to Mozambique, but this propulsive 2006 Flatts hit makes the journey a quick one—and in the process, a broken relationship is mended by sharing the road together. The music is catchy and the line you will remember is just four words—Life Is a Highway. It pulls you right in. The best hooks are easy to say and easy to remember. Brevity is powerful.

What are your "hooks" when communicating? While they may not be as colorful as Rascal Flatts', your hook can be just as effective. What theme or action do you want them to focus on? Summarize it in a few simple, repeatable words and use it over and over. Jim Collins did this when he titled his book *Good to Great*. It takes effort to find just the right phrase, but when you nail it, it's yours and it pays.

8.3.3   Simplicity

If it takes a lot of words, complicated slides, or too much explaining, you're lost before you started. In "Could I Have This Dance," Anne Murray sings, "Could I have this dance for the rest of my life?/Would you be my partner every night?" Don't these simple phrases just make sense? It's the basics of life summed up in a few easy phrases. How do you use simplicity to your advantage in business? Focus on the essentials; pare out extra words and phrases; nix the unnecessary jargon; talk like a human being. Think about how you like to be communicated with and use that to reframe your own efforts.

Country music is simple at its very foundations. It's a wise policy. The simple approach was good enough for Strunk and White, who penned an entire chapter in their book, *Elements of Style*, which reads "omit needless words." It's probably good enough for you, too. Say what you came to say. Say it passionately. Then get out of the way.

8.3.4   Speak from the heart

Perhaps the biggest draw of country music is the emotional connection of talking straight from the heart. This isn't for every situation in business communication, but there are times when a heartfelt message is appropriate and meaningful. Be open to sharing your vulnerable side and show people your humanity and empathy for others. This builds respect and invites people to listen. They see you are human and are treating them the same way. For this example, in Brad Paisley's song, we can find the following love lines: "I can just see you with a baby on the way/I can just see you when your hair is turning gray/What I can't see is how I'm ever gonna love you more/But I've said that before." It's honest, tells a story and it's clearly from the heart. Paisley wrote this for wife Kimberly after the birth of their son Jasper, to show how his love has grown for her every day. "The day before [the baby was born], I thought I loved her," he explains. "But then 24 hours later..."

Be genuine and you can strive for this genuine tone in your business communication. To copy this approach, aim to speak the truth about who you are, what you do, and perhaps most importantly, why you love to do it.

Country music is primarily values based. The list goes something like: God, family, country, hard work, and getting over heartache. Even if you have a completely different set of values, you can't help but admire the straightforward way that these are presented. In your own communication, you can do well by trying to emulate this approach. Potential customers can likely get your product or service from numerous other sources, and want to know what makes you different. Speaking to your values, and giving them a chance to form a connection, is a great way to compel someone to shop with you.

Do you want to up your business communication skills? Country music may not be your thing, but there's a reason it's so popular. Challenge yourself by trying some of these tips. Take something you want to communicate and use one or two of these approaches. See how your communication changes when you simplify, be human, or tell a story. Watch how people respond when you and your audience connect.

8.4 *"Action speaks louder than words"—movement and gesture in business communication*

"Action speaks louder than words." In business communication, "Body language is a broad term of communication using body movements or gestures..." Movement in fact is a main source of communication for modern human beings and is the earliest and most primitive as well as the most advanced form of communication. Even spoken language is based on movement and gesture. That is, in order to speak, words must be formed, articulated, and expressed via a complex synergy of movement involving not only the lips, larynx, lungs, and tongue, and via complex programming which takes place in the cerebellum, basal ganglia, and motor neocortex of the brain.

The book titled *What Every Engineer Should Know About Business Communication*, states:

> Hand gestures are important to emphasize words and emotions, illustrate verbal messages, or even replace verbal messages altogether. Use gestures to paint a picture for the audience or underline key phrases.

Gestures can be executed more rapidly than speech, and can convey concepts that are quite cumbersome to describe verbally. Indeed, it has been estimated that the human hand is 20,000 times more versatile than the mouth in producing comprehensible gestures.

Angela Isadora Duncan (1877–1927), an American dancer, lamented:

> Very little is known in our day of the magic which resides in movement, and the potency of certain gestures. The number of physical movements that most people make through life is extremely limited. Having stifled and disciplined their movements in the first stages of childhood, they resort to a set of habits seldom varied. So too, their mental activities respond to set formulas, often repeated. With this repetition of physical and mental movements, they limit their expression until they become like actors who each night play the same role.

Among humans, it is the right half of the brain, coupled with limbic influences and that of the basal ganglia which provides and extracts emotional meaning from movement. Over the course of human evolution, neuronal organization has become more complex and new brain structures have evolved that have made possible the translation of movement into meaningful gestures and then, finally, into complex human speech. In fact, it appears that the development of complex gesturing was a precursor to the development of spoken language, and that, conversely, spoken language required that movement be modified and imposed on sounds. Before the first word, there was the sign, and the sign still accompanies and gives meaning to the spoken word, even when nothing has been said.

8.5   Business communication: Never underestimate our ability to persuade with our eye

The eyes communicate powerful cognitive messages in business communication and human interaction for group technology in industrial design engineering.

Why eye contact? There's an old myth if you won't look at me I can't trust you. It might be true, might not be. But if they believe it, it's true! The eyes are one of the most important nonverbal channels you have for communication and connecting with other people.

Eyes are not only the "window to the soul"; they also answer the critical question when you are trying to connect: Is he or she paying attention to what I'm saying?

Not only is direct eye contact a must for effective communication, it's one of the qualities of being an effective leader. The two—effective communication and leadership skills—walk hand-in-hand.

It's true with adults, and it's true with babies. According to a 1996 Canadian study of 3- to 6-month-old infants, smiling in infants decreased when adult eye contact was removed. We respond to eye contact because it confers meaning. In the stories associated with the 10th anniversary of the September 11th attacks, one firefighter wrote of the importance he learned to ensure he made direct eye contact each morning with his family, not knowing if it might be their last.

Let's get back to our childhood. When a mother stares into her baby's eyes, the baby's oxytocin levels rise; this causes the infant to stare back into its mother's eyes, which, in turn, causes the mother to release more oxytocin, and so on. This positive feedback loop seems to create a strong emotional bond between mother and child during a time when the baby can't express itself in other ways.

As shown in Figure 8.3, in business communication, we should never underestimate our ability to persuade with our eye. The importance of eye contact for communication and interaction can be demonstrated by research, which shows that when our canine pals stare into our eyes, they activate the same hormonal response that bonds us to human infants. People and dogs often look into each other's eyes while interacting—a sign of understanding and affection.

Here, mutual gazing has a profound effect on both the dogs and their owners. During a research experiment, of the duos that had spent the greatest amount of time looking into each other's eyes, both male and female dogs experienced a 130% rise in oxytocin levels, and both male and female owners a 300% increase.

Figure 8.3 Business communication: Never underestimate our ability to persuade with our eyes.

The experiment results above suggest that human–dog interactions including eye contact elicit the same type of oxytocin positive feedback loop as seen between mothers and their infants.

Eye contact generates oxytocin, a hormone that plays a role in gaining trust. Since trust is the foundation of effective business communication, we should never underestimate our ability to persuade with our eyes.

A growing body of research indicates that eye gaze denotes status. Whether in dyads or small groups, those with status receive more eye contact from their conversation companions. Nonverbal communications include tone of voice, facial expressions, gestures, and body language, including posture, and of course: eye gaze. At least 65% of communication is nonverbal. If more people understood this, they might spend more energy on how they speak, not just on which words they choose when speaking.

A person who stands up straight and looks you in the eye while speaking will get more attention and respect, in general, than a speaker who slouches and has difficulty maintaining eye contact. The two speakers may use the exact same words, and even have an equal understanding of the subject on which they speak. But the one who uses nonverbal communication techniques well will seem to know more than the other.

The study of eye contact is sometimes known as oculesics. Although eye contact and facial expressions are often linked together, the eyes could transmit a message of their own.

Eye contact is a type of nonverbal communication that is strongly influenced by social behavior. In the Western civilizations, eye contact is most often defined as a sign of confidence. Eye contact is not consistent among different religions, cultures, and social backgrounds.

8.5.1   Cultural differences

"Language is more than just words," wrote Alix Henley and Judith Schott. In business, one must be aware of cultural differences. In some cultures, looking people in the eye is assumed to indicate honesty and straight-forwardness; in others, it is seen as challenging and rude. Eyelids are a barrier, and using that barrier to restrict eye contact can be perceived in different ways depending on the culture.

- In Western societies, eye contact can imply empathy and comes across as a type of emotional connection.
- In the United States, the practice is often "eyelids up." You want to make eye contact while meeting someone, listening to a person, and saying good-bye. It supports your ability to forge a relationship. In the United States, eye contact indicates a degree of attention or inter-est, influences attitude change or persuasion, regulates interaction, communicates emotion, defines power and status, and has a central role in managing impressions of others. That guidance holds true for Western nations in general.
- In English culture, a certain amount of eye contact is required, but too much makes many people uncomfortable. Most English people make eye contact at the beginning and then let their gaze drift to the side periodically to avoid "staring the other person out."
- In the nineteenth century, Spanish women used eye contact to say what they couldn't express explicitly.
- In Asia (including Middle East), keeping eye contact with someone of authority implies rudeness and can be mistaken as a provoking means of communication. In some Asian countries, avoiding eye contact is a sign of respect. In South Asian and many other cul-tures, direct eye contact is generally regarded as aggressive and rude.
- In Middle Eastern countries, men tend to lock eyes with each other as a way of conveying, "I mean what I say, and what I say is the truth." Inter-gender eye contact is not the same. In fact, if you are a woman doing business with a man from the Middle East, try to ascertain how much contact he has had with Western practices. Use the barrier of your eyelids to put psychological distance between you and the individual; it's often seen as respectful. Most people in Arab cultures share a great deal of eye contact and may regard too little as disrespectful. Arabic cultures make prolonged eye contact—they believe it shows interest and helps them understand the truthfulness of the other person. A person who doesn't recipro-cate is seen as untrustworthy.
- Japan, Africa, Latin American, and the Caribbean—avoid eye con-tact to show respect.

Different cultures also vary in the amount that it is acceptable to watch other people. Some experts call these high-look and low-look cultures. British culture is a low-look culture. Watching other people, especially strangers, is regarded as intrusive. People who are caught "staring" usually look away quickly and are often embarrassed. Those being watched may feel threatened and insulted. In high-look cultures, for example, in southern Europe, looking or gazing at other people is perfectly acceptable; being watched is not a problem. When people's expectations and interpretations clash, irritation and misunderstandings can arise.

In many other countries, eyelids are a tool of maintaining a polite and confident demeanor. Make eye contact periodically, such as when meeting a person, making or listening to an important point, and saying good-bye. Check local practices through observation and by asking colleagues in the field.

Eye contact can indicate how interested a person is in the communication taking place. It could also suggest trust and truthfulness. Often, when people are being untruthful, they tend to look away and resist eye contact.

Furthermore, eye contact portrays someone's involvement and attention. Attention is a function of eye contact that can be both negatively and positively affected by a person's gaze. The latter can show confidence, anger, fear, and so on.

A person's direction of gaze is important. To engage in a productive communicative session, attention must be given. Looking away often demonstrates a lack of involvement in the conversation. Autism is characterized by impaired social interaction and communication. A lack of eye contact is not a cause of this disease, but can often be a sign of its presence. This lack of eye contact provides information on how the individual's lack of attention can lead to a lack of communication skills. People suffering from social anxiety (or social phobia) also resist eye contact, although this does not necessarily mean that there is a lack of attention.

8.5.2   Monologue or dialogue

If you are giving a presentation, look at your audience—as individuals. Even if you are addressing an audience of thousands, pick a spot in the audience and direct your eyes to it. Deliver your insight or instruction, and then move to another spot and make eye contact. You want to convey a sense of focus and interest in each member of your audience, so having your eyes dart around the room will not help you do that. Neither will reading your notes and making no attempt to connect visually. Both eye contact errors will probably cause you to lose your audience.

If you are engaged in a dialogue, follow the cultural guidance in Section 8.5.1, as well as the guidance in Section 8.5.3.

8.5.3 Nature of the conversation

As a business professional, one of the most important skills you can ever cultivate is active listening. That means you listen with your entire body, from eyes to toes. Since we're just focused on eyes here, an important element of active listening is looking at the person talking. Never "leave" a conversation by checking for text messages, or even for a split second glancing at your laptop to see if a new email has come in. Never look out the window at the gray squirrel running through the parking lot while someone is talking to you.

In the business world, establishing eye contact shows confidence. This confidence leads to maintaining a respectable, authoritative, professional, and competent allure. However, strong eye contact could portray someone's arrogance, overconfidence, and could be seen as aggression.

Now switch your point of view. You are the one talking, and the other person is listening. You ask a question and the person's eyes drift off to the left or right. What do you make of that? This is a normal way that people respond when they are processing information. There is nothing inattentive or rude about it. When the person arrives at an answer or has a comment, he or she will likely make eye contact with you again.

8.5.4 Risk engineering of nonverbal business communication

Eye contact is normal. Eye contact is natural. Eye contact is healthy. Infants instinctually have a strong desire to gaze into others' eyes. Yet, why is it that so many adults with fairly good interpersonal communication skills have long since "unlearned" this fundamental and indispensable nonverbal behavior?

In the science of nonverbal communication, the most crucial portion of the body is the face. And the most important part of the face is the eyes—the eyelids, the eyebrows, and the regions around the eyes.

When it comes to body language, people often ask, "Does good eye contact mean I have to look the other person directly in the eye all the time?" (or some variation of it). The short answer is definitely "No." Here's the long answer: If you stare directly into one eye of a person—or switch back and forth between his or her eyes, it quickly becomes too psychologically intense. It is almost always interpreted (depending on the other signals and the context) as predatory behavior, anger, sexual attraction, or deception.

If you're participating in healthy conversation (and not experiencing the above emotions), "eye contact" is effectively defined as looking semi-randomly in an area whose borders surround the eyes by about 2 cm. This would be between 30 and 70% of the time. Although they're not in this "eye contact ellipse" (ECE), it is important not to stare at the forehead (lest you intimidate) or at the mouth (as this will send signals of sexual interest). The reason that I said "semi-randomly" is that you should briefly fixate on one of the person's eyes. You should then quickly (and very briefly) fixate on some other portion of the person's face within (and sometimes outside) of the ECE. You should then return to the same or opposite eye. (Remember, it is important to *not* gaze within the ECE the balance of the time.) These rapid eye movements or saccades occur naturally and subconsciously during healthy conversation. They also occur other times such as during reading, thinking, and so on. Staring occurs when eye movements and blinking are greatly diminished or absent. We are not aware of these patterns of eyelid movements the vast majority of the time—although sometimes they occur at the edge of consciousness. There are occasions though for different reasons that these do become conscious acts. With rare exception, a person should never fixate on one area too long (2 seconds is a good limit) or too intensely (with no blinking or eye movements).

Defining "eye contact" as sometimes not looking directly into someone's eyes, but close to them, may seem somewhat counterintuitive. But this is what we do during healthy conversation. Generally the speaker has a natural decrease of this "eye contact"—closer to 30%; the listener, on the other hand, experiences an eye contact crescendo—closer to 70%, most of the time. If we want to build and engender rapport, we need to be aware of and avoid this tendency to decrease eye contact when our role changes to speaker.

As discussed earlier, when a person tells a lie, his or her eye contact will often increase to what most people consider "staring." This is a great example of an overcompensating behavior. It is also true that the opposite may occur—a dramatic drop-off of eye contact during lying. It is important to compare the amount and pattern of eye contact in a dynamic context—contrasting the times before and after the suspected moments of deception. If a lie is suspected, the specific subject matter should be revisited. Using the Socratic Method and noting whether similar patterns of nonverbal behavior are displayed will help in validating or refuting deception. This is only one of many nonverbal techniques used in detecting deception.

Although an example of an exception, in certain cultures a lack of eye contact is considered a sign of respect. Yet in many countries, very little or no eye contact during an encounter is a signal of extreme disrespect. Sometimes, it is an effort to avoid an escalation of negative emotions.

Another example where eye contact is very crucial, yet under delivered, is during a handshake. It's amazing how many "professionals" lack this important component of the greeting. A helpful technique is to observe the color of the other person's eyes (irises) during the handshake. And while noting this, repeat a positive and sincere mantra, silently to yourself. This may sound trite or mawkish, but it works. I've been practicing this since I've been a teenager. Doing so has the ability to bring you closer and to build true rapport during the critical and vastly underappreciated few seconds of a handshake/greeting.

Very few of us are aware of just how important the eyes are when it comes to smiling. One requirement for a smile to be sincere is there must be a dynamic and momentary partial closure of the eyelids. But it is mainly a secondary, passive closing. Without a partial eyelid closure, the smile is insincere—no exceptions. This does not mean the person as a whole is insincere—merely in that moment, regardless of what the mouth looks like or what words are being spoken—they are "pushing out" what is known as a "social smile." They are acting happy (or happier) than their true emotions—in that moment.

Since the eyes are the only part of your central nervous system that makes contact with the outside of your body, the old adage of "The eyes are the windows to the soul" has basis in medical fact. The practice and study of eye contact is just one aspect of the immense nonverbal value of this most precious and mysterious organ. Ignore it at your own risk.

8.5.5   Eye contact, eye communication and eye roll

Eye contact is fleeting. It can be in passing, just a glance or a fraction of a second. It can be eyes flitting across an audience not really connecting. It could turn into a form of connection, but it isn't necessarily communication.

Eye communication is connection—think of eye contact on steroids. It's the act of two pairs of eyes connecting and the contact leading to communication. Eye communication involves more extended eye contact (at least 3–5 seconds for speakers communicating to a group) that forms a bond between two people. As a speaker communicating to an audience, eye communication is the key to engaging with your audience. It makes your presentation more like a conversation than a pitch.

Eye communication:

- Establishes rapport
- Strengthens listener involvement
- Contributes to a higher retention level of eye contact
- Increases your ability to persuade

A successful communications experience requires engaged eye communication. If the listeners' ears are open but their eyes are closed, no connection can be made. Eye communication is the key.

On the other hand, eye roll is discounting what the other person said. It is a put-down, and is actually much stronger a put-down than we tend to think. Often after an eye roll is picked off by the other party, we hear the reply, "But I didn't say anything."

"To make oneself understood to the people, one must first speak to their eyes," said Napoleon Bonaparte.

Don't just make contact; communicate with your eyes, positively.

8.5.6   Three easy tips for using eye contact for better communication

Eye contact may seem like a natural behavior that needs no instruction. But the way in which you make eye contact while communicating can have a significant impact on how others perceive you. Take a look at our three tips below for using eye contact to come across as an approachable, confident professional.

1. One of the trickiest parts about effective eye contact is deciding how long to hold it. Maintaining eye contact for too long can feel aggressive and make your conversation partner uncomfortable. Not making enough eye contact will make you appear nervous or disinterested. In order to appear confident, engaged, and approachable, maintain eye contact throughout a conversation, but every few sentences, shift your gaze for about a second before resuming eye contact. You can also shift your gaze from one of the speaker's eyes to the other in order to avoid looking as though you are staring too intensely.
2. When breaking eye contact, be mindful of where you look; the direction of your gaze can impact the way your listener perceives you. Looking downward typically conveys a lack of confidence or disinterest in the conversation. Looking upward can make you appear thoughtful, but can also be slightly distracting. Looking to the side is usually your best bet; this breaks the intensity of extended eye contact, but in a way that is so subtle, your listener will likely not even notice.
3. When giving a presentation, use eye contact to your advantage to engage the entire audience. It's common for people to make eye contact with only the listeners in the front of the group or to make repeated eye contact with someone they know and are comfortable with. People are much more likely to maintain interest and remember the message if the speaker makes direct eye contact. Throughout your presentation, make eye contact with individual members of the audience for 3–5 seconds each, varying the sections of the audience that you are looking at in order to draw in the entire group.

"The cheapest, most effective way to connect with others is to look them in the eye," said Nicholas Boothman. So, concentrate—don't start dreaming—and keep eye contact.

8.5.7   Eye cues

The eyes can give valuable clues about how a person thinks. People have different ways of communicating their experiences:

- Some express themselves in pictures,
- Others talk about how things sound to them, and
- Others speak about how things feel.

People have different mental maps which drive their behavior. Here, a mental map is a powerful way of expressing the thought patterns, pictures, and associations that already exist in the brain. When new information is compatible with your knowledge structures, it is accepted; when it does not mesh with your preconceived ideas or past experience, it receives little consideration, is distorted or ignored. "We see the things not as they are, but as we are," said H. M. Tomlinson.

Kinesthetic people will tend to look down more, while visuals spend more time looking up, and auditories look sideways. "This is because they each favor one sense to code and store general information as well as express it," writes Nicholas Boothman. "If you asked, 'How was the Stones concert?' a visual would first remember how it looked, an auditory how it sounded, and a kinesthetic how it felt. But eye cues can tell you more than who you're dealing with; they can also tell you what you're dealing with." When people look up and right, they are probably constructing, or making up, their answer. When they look up and left, they are more than likely remembering it. We can use eye contact to validate business—and personal—connections.

Generally, direct eye contact confers attention and respect. It says much, without saying a word... in your professional life, as in your personal life.

8.6   Teleportation: Can I leave my message with the moon?

Mid-autumn night, my son Cheney asked me: "Dad, does teleportation really exist?"

Teleportation is the theoretical transfer of matter, energy, or data from one point to another without traversing the physical space between them.

Scientists believe we could then teleport to the moon or even Mars by the end of the century.

The key to teleportation is to use the entangled pair as a communication channel to transmit the information you want to teleport from one place to another. The exact teleportation protocol is a little tricky, so bear with me for a moment. Let's imagine you wanted to teleport a particle A from the Earth to the moon. First of all, you would need to generate a pair of entangled particles B and C. One of those (B), you would keep here on Earth, while the other one (C) would be sent to the moon at the mid-autumn night.

The objective of quantum teleportation is to change the identity of the particle C on the moon and to turn it into an exact replica of A. In order to do this, we somehow need to extract all the information about A, and teleport that information to the moon, so that C can use it as a kind of blueprint to turn itself into A. The problem however is that we cannot simply scan A to extract its information.

$$\text{Earth: } A - B - - - - - - - - - - - - - - - - - - C\text{: Moon}$$

So instead, we entangle A with the other particle B here on Earth. That way, some of A's information gets shared with B, and since B is entangled with C on the moon, the information is passed on further to C, enabling C to transform itself into a copy of A.

There is a price to pay, however: the original particle A will be destroyed in the process as it loses all its information and thus also its identity. The same also happens to B. In summary, you would see the original particle A disappear here on Earth, and it appears again an instant later on the moon. Here, you have successfully teleported your first particle.

With teleportation, you can transmit quantum information to the Moon instantaneously. This reminds me of an ancient Chinese poem which includes such a sentence: I leave my message with the moon.

Gazing at the Moon, Longing from Afar

The moment that moon grows full, the sea gives birth to a
  shining moon.
Brightening by heaven's light, the other end of the world shares
  this moment.
Separated by the lengthy night, the whole evening gives rise to
  longing.
Extinguishing a candle, I feel fullness of moonlight.
Awaking in heavy dew, I put on cloths.

Unable to deliver in writing, I leave my message with the moon.
Returning to bed, I fall into a beautiful dream of reunion.

Jiuling Zhang (678–740)

Yes, it's the moon that engineers a great business communication with the people we wish to connect with, especially at the mid-autumn night.

8.7 Reflection on 4th: Collaborative action and pervasive communication

Collaborative action and pervasive communication are critical to implement group technology in industrial design engineering. Reflecting on the 4th of July, we can see that the American Revolution is different from other revolutions in that:

- The American Revolution was not characterized by guillotines or extensive bloodletting. "It did not devour its own children, as most revolutions do." Instead, it was characterized by communication/argumentation that seeks to persuade others through reasoned judgment.
- The American Revolution was a collective achievement of numerous parties and groups that eschewed a single figurehead. To a great extent, group technology was utilized to justify actions and beliefs and to influence the thought and actions of other parties and groups.
- Despite vast differences among the numerous parties and groups, the American revolutionaries chose to argue with each other, applying argumentation as a tool of decision-making under uncertainty.

Through collaborative action and pervasive communication, the American Revolution achieved three things:

- The successful war for colonial independence.
- The first enduring, large-scale Republic in the world.
- The first nation built on liberty.

"What is the difference between liberty and freedom?" Once I asked a friend who has served in the US Navy.
"Liberty is the action to attain freedom," the friend said.
Yes, collaborative action and pervasive communication is the key to attain and defend freedom.

8.8   Alice and Cheshire Cat: Risk engineering of our communication

In honor of the 150th birthday of *Alice's Adventures in Wonderland*, events are going on throughout the world. *Alice's Adventures in Wonderland* is an 1865 novel written by English author Charles Lutwidge Dodgson under the pseudonym Lewis Carroll. It tells of a girl named Alice falling through a rabbit hole into a fantasy world populated by peculiar, anthropomorphic creatures. The tale plays with logic, a tool to mitigate the risk of fallacy in communication.

One day in 1862, often said to be July 4, Charles Lutwidge Dodgson and the Reverend Robinson Duckworth rowed in a boat, up the River Isis with the three young daughters of Henry Liddell (the Vice-Chancellor of Oxford University and Dean of Christ Church): Lorina Charlotte Liddell (aged 13); Alice Pleasance Liddell (aged 10); and Edith Mary Liddell (aged 8). While rowing, Dodgson told Alice and her sisters a tale of a little girl who, while half-dozing on the riverbank with her sister, suddenly saw a white rabbit run past her with a pocket watch and followed him right into a rabbit hole to Wonderland. The girls loved it, and Alice begged him to write the story down for her. Three years later, copies of it hit book-stores. The tale plays with logic, giving the story lasting popularity with adults as well as with children.

As a mathematical professor, Charles Lutwidge Dodgson, aka Lewis Carroll, fused his passion for logic, mathematics, and games with his love of words and nonsense stories to produce a multifaceted, intricately struc-tured work of literature, revealing Dodgson's profound knowledge of the rules of clear thinking, informal and formal logic, symbolic logic, and human nature.

For example, as shown in Figure 8.4, Alice has a logical conversation with the Cheshire Cat, who assures her that they are both mad. She does not accept his proof of her madness ("You must be, or you wouldn't have come here"), but asks how he knows that he is mad:

"To begin with," said the Cat, "a dog's not mad. You grant that?"

"I suppose so," said Alice.

"Well, then," the Cat went on, "you see a dog growls when it's angry, and wags its tail when it's pleased. Now I growl when I'm pleased, and wag my tail when I'm angry. Therefore I'm mad" (Chapter 6).

In his argument, the Cat commits the logical fallacy of denying the antecedent:

If an animal growls when angry and wags its tail when pleased, it is not mad (If P, then Q).

I growl when pleased, and wag my tail when angry (Not P).

Therefore, I am mad (Therefore, not Q).

Figure 8.4 Alice speaks to Cheshire Cat. (Courtesy of Library of Congress, https:// blogs.loc.gov/loc/files/2016/05/20160330SM031-184x300.jpg.)

This is an invalid argument because, even if the premises are true, the argument does not sufficiently prove that the Cat is mad.

Alice's Adventures in Wonderland, far from being just an entertaining children's book, is more profound and deeply reflective of Dodgson's character than it may seem. It reveals a common yet understudied aspect of communication: mitigate the risk of fallacy.

8.9   Vocal quality in business communication and timbre in Claude Debussy's La Mer

8.9.1   Vocal quality and timbre

"Studies show that 38% of the impact of our presentations comes from what the audience hears—not the words, but the vocal quality, volume, pace, and expression we project."

Timbre is the quality of a musical note or sound or tone that distinguishes different types of sound production, such as voices or musical instruments. Timbre is also known as tone quality or tone color.

This section identifies the basic categories of timbre employed by Debussy in La Mer. We will describe how timbre (vocal quality) is employed structurally through the course of the music movement, with clear identification of structural points and relationships between the sections. We are going to discuss how timbre in this first movement is fundamental compared to other parameters. La Mer was unlike any other piece composed at the time. Claude Debussy utilized timbre as relative to the structure of La Mer.

8.9.2   La Mer and timbre

One hundred and nine years ago, classical music witnessed a sea change—quite literally. On October 15, 1905, French composer Claude Debussy's symphonic portrait of the sea, called "La Mer," premiered in Paris. The way Debussy captured the ocean's color, light, and mood—using the orchestra as his paintbrush—gave composers new ways to think about writing orchestral music. With "La Mer," Debussy ignored the old rules about combining textures and sounds in symphony orchestra, and created a whole new world of sonic possibilities. It's an incredible experience: an atmosphere of nature-inspired shades, moods, and colors. These sensual qualities are what Debussy's music is all about, and La Mer is a perfect example.

Debussy's use of timbre is fundamental throughout La Mer and perhaps is more crucial than other parameters that other composers might usually rely on, such as tonality and structure. However, for Debussy, the use of timbre is very important when reproducing and replicating the sounds of the great ocean. The timbres he has used throughout reflect the title of the first movement and do suggest characteristic features one might associate with the sea.

Debussy's sophisticated use of timbre is shown in his orchestral scoring of the first movement of La Mer. The use of timbre and effect it has in Debussy's orchestration not only becomes an important part of his formal structure, but they also provide energetic pictorial images and emotional atmospheres demonstrating his close relationship of music with visual impressionism. Although Debussy's usage of instruments and most of his instrumentations are not as aggressive as his fellow composers such as Mahler, Strauss, and Stravinsky, his delicate way of exploiting timbre is one of his most important contributions.

Debussy utilizes timbre (vocal quality) to suggest color, mood, and atmosphere, as would Monet or Renoir in their own paintings. Though the work is presented as a series of three sketches, it has a distinctly symphonic construction; it's forceful, larger-scale outer movements contrast with the playful middle movement that functions like a scherzo. Debussy gave pictorial names to each sketch, but he warned that they were not meant to be taken literally (although his friend Erik Satie joked that he liked "From Dawn to Noon on the Sea" very much, "especially the part from 10:30 to a quarter to 11:00"). Yet the work certainly conjures images of the sea, from gentle swells and sparkling surfaces to the crashing of breakers on the shore.

8.9.3 *Debussy by the sea*

Debussy began writing it in 1903 in Burgundy, which actually is not near the sea. He sketched La Mer in 1903, and in 1905 when he traveled to Eastbourne, a town on the English Channel, to complete the piece at the seaside. In La Mer, Debussy composed his musical portrait of the sea. He began working on it in the town of Bichain which is actually far inland, perhaps 100 miles southeast of Paris toward Switzerland, in the historic region of Burgundy. But much of the time he was working on it, he was staying in Pourville. His purpose, as he later wrote to his stepson, was to depict the ocean's constant mutability in ways that the painters in his Impressionist milieu could not. "Music," he wrote, "has this over painting: it can bring together all manner of variations of color and light." Finishing it March 1905, he spent the month of August on the English side of the Channel, at Eastbourne, and on August 7 he was correcting the publisher's proofs in advance of the October premiere in Paris.

8.9.4 *Debussy and poetry*

Debussy loved the poetry of Baudelaire, as he loved the painters Turner, Hokusai, and Monet, all of whom created memorable images of the sea. When he published the score of his orchestral work, La Mer, he requested that one of Hokusai's prints, "The Hollow of the Wave off Kanagawa," be reproduced as part of the cover design. "I am working on three symphonic sketches under the title La Mer....You may not know that I was supposed to have been a sailor, and only by chance did fate lead me in another direction. But I have always retained a passionate love for the sea."

Debussy's La Mer is a unique mix of tone poem and symphony, a three-movement impression of the ocean. Debussy was decidedly not interested in pictorialism or program music. Indeed, only a few months earlier, in a concert review for a Paris newspaper, he had written that the popularity of Beethoven's "Pastoral" Symphony "rests on the common and mutual misunderstanding that exists between man and nature." Debussy wrote that the bird calls in that symphony were "more like the art of M. de Vaucauson [an eighteenth-century creator of a famous mechanical duck] than drawn from nature's book. All such imitations are in the end useless—purely arbitrary interpretations." In other places, he wrote, the "Pastoral" succeeded "simply because there is no attempt at direct imitation, but rather at capturing the invisible sentiments of nature."

In La Mer, a vivid landscape is suggested from the various wave figurations, shimmering light, and onomatopoeic sound effects. The vagueness, ambiguity, and effects of light he uses reflects the visual arts, and are vividly implied from his usage of instrumental echoing effects, tremolo strings, harp, and the special timbre of percussion instruments.

8.9.5   Debussy and impressionism

Debussy was heavily related to impressionism, which was dominated by atmosphere and the use of suggestion. Debussy was a very visually oriented composer. Many of his works are small musical miniatures with evocative titles—think of "Clair de Lune" (Moonlight) or "Girl with the Flaxen Hair." In fact, there are series of short works simply called "Images." His studio was full of prints of paintings or those postcard-like souvenirs one might find at a museum—images which, given the vagueness of his harmonic style and almost anti-melodic approach to sound, earned him the title "Impressionist."

Usually, we tend to think of "Impressionism" in painting as soft and flexible, playing more with light than substance. This is easy to induce musically by the use of non-traditional scales, especially the whole-tone scale which has no harmonic function we associate with tonality, especially the strong functions of chord progressions like the dominant to the tonic resolution that gives it a satisfying, structural coherence. In several works by Debussy—think "Prelude to the Afternoon of a Faun" or, again, "Clair de Lune"—the harmonic vagueness is matched by softer dynamics and even though there are climaxes, they are almost understated.

Debussy called La Mer "three symphonic sketches," avoiding the loaded term symphony. Yet the work is sometimes called a symphony; it consists of two powerful outer movements framing a lighter, faster piece which acts as a type of scherzo. The author Jean Barraque (in "La Mer de Debussy," *Analyse musicale* 12/3, June 1988), describes La Mer as the first work to have an "open" form—a *devenir sonore* or "sonorous becoming... a developmental process in which the very notions of exposition and development coexist in an uninterrupted burst." He was also influenced by the "infinite arabesques" and complex counterpoint of the Javanese gamelan, a unique and exotic sound-world he first heard in 1889 at the Universal Exposition in Paris.

8.10   Debussy and Japanese artist Hokusai's painting "The Great Wave of Kanagawa"

As shown in Figure 8.5, one of Debussy's greatest influences when composing La Mer was the Japanese artist Hokusai's painting "The Great Wave of Kanagawa," which displays a vivid but suggestive, powerful wave breaking with foam and spray crashing, creating a scene of risk.

Figure 8.5 Japanese artist Hokusai's painting "The Great Wave of Kanagawa." (Courtesy of Library of Congress.)

Debussy was just as influenced by the stylization of nature as seen in the landscape prints from the "The Great Wave of Kanagawa," which he had in his studio and which adorned the first printed edition of Debussy's score.

"The Great Wave of Kanagawa" is very similar to Debussy's first movement, which is vague but also has moments of terror in it. This can be seen at bar 84. Here the new section starts. In the background to this, the second violins and violas wave-like ostinato figure suggests the shimmering, repetitive surface of the sea, while the cellos' wider range and more active motion might begin to imply more movement of the sea below the surface. In this section, the harps also play rolling chords that add resonance and suggest the relentless rays of the sun.

8.10.1   La Mer's structure

The Debussy La Mer has the subtitle *Trois esquisses symphoniques* ("Three symphonic sketches"). However, Debussy didn't give any kind of symphonic structure to La Mer. It's more of a series of spontaneous fragments of music hung together to create an exquisite and mysterious world of beauty. Debussy actually ignored pretty much all of the old rules about proper music-writing in La Mer. He used his own glowing harmonies which broke basic music tradition, and used the orchestra as a breathtaking canvas of luxurious nuances.

La Mer is split into three different movements, the first being "From Dawn to Noon on the Sea" with "quick timbral changes to suggest the sea's different, ever changing natures." "From Dawn to Noon on the Sea" is built on short episodes, which use different instruments to suggest the various timbres of the sea. Debussy develops this to a "wonderful sugges- tion of the swelling of waves, as a theme for divided cellos swells and sub- sides"; this is similarly copied by the "timpani and horns." These episodes can also be heard in the final stages of the final movement.

The first movement is entitled "From Dawn to Noon on the Sea," and the final movement is the "Dialogue of the Wind and the Sea." These are comparable to the substantial outer movements one might find in a sym- phony. The middle movement is a light, scherzo-like movement, almost a waltz, entitled "Play of the Waves."

8.10.2   Movement 1. De l'aube a midi sur la mer ("From Dawn to Noon on the Sea") (B minor)

In the first movement of La Mer, Debussy's remarkable use of timbre can be seen. As the sun slowly rises, at bars 1–5 the violas, doubled at the octave, add their ascending pentatonic melody, which gives sense of a fresh and clean start to the day. At bar 31 the first principal section follows. This part of the movement grows and develops its own material, being mostly independent. However, we can see it is still based on the opening, with a few different elements omitted. For instance at bar 33, we can see that the flute's pentatonic motif is doubled by the clarinets at the octave below. The use of the pentatonic melody might suggest an oriental flavor. The use of the flutes that dominate the hollow consecutive fifth have a similar timbre to that of the Chinese flute. Debussy gives a pictorial title for his first movement, "From Dawn to Noon on the Sea." Through the first movement we can hear the gentle swells and glistening surfaces of the sea to the breaking of the waves on the shoreline. Perhaps through the quieter parts of the movement it suggests that under the sea it is calm and unchanging yet on the surface is an ever-changing picture.

8.10.3   Movement 2. Jeux de vagues ("Play of the Waves")

Allegro (C sharp minor)

"Play of the Waves" is an interlude of sprightly waves in the form of different instruments darting about on a summer day. Its three beats to the bar gives it a feeling of a lively waltz. The second movement was play- ful and is powerful and urgent. The brass theme from the opening of the first movement is heard again, now a firm statement allowed to develop and grow within the larger dialogue, and contrasted against a theme of falling triplets that sound by turns either vigorous or languid. The second movement is livelier, and uses repeated orchestral swells with just about every combination of instrument you can think of. Again, the listener's imagination can roam around the mysterious textures. The full orchestra gathers its forces for a triumphant conclusion, timpani and bass drum accentuating the power of Debussy's majestic sea.

8.10.4   Movement 3. Dialogue du vent et de la mer ("Dialogue of the Wind and the Sea")

"Dialogue of the Wind and the Sea" summons God and man, and with them an air of menace and space. Its orchestration, more overt and powerful than in the first movement, helped dispel the notion that Debussy's music was, as expressed at the time, effeminate and decadent. It recalls two themes from the first movement but is less a narrative and more a tableau of scenes and effects. The opening is a rush to the surface in the cellos, their progress halted by a stern English horn. A theme from the first movement returns but sounds more ominous. The sea is wilder, tougher, and stormier now, as well as darker and more brisk and mercurial in its moods. The whole-tone scale is more prominent, as are many ocean effects, for example, deep, globular sounds in the low strings, skipping waves in the violins, and so on. A broad theme sings in the woodwinds over heaving strings. After a huge climax, trumpets, horns, and then trombones tumble and disappear into the waves. Calm returns briefly before the sea becomes frenetic with skittish rhythms in the trumpets and wild calls from muted cornets. The broad woodwind theme returns and evolves into the horn chorale from the first movement, now sung powerfully by the trombones, leading to a glorious climax. "Dialogue of the Wind and the Sea" is the most dramatic movement. As the title hints, the music is about the forces of wind and ocean clashing. The movement is animated and tumultuous.

8.10.5   La Mer's influence

La Mer has been influential on many contemporary soundtrack composers because of its highly suggestive and moody atmosphere. For example, some of La Mer's passages (the third movement, for example) may have inspired John Williams for the score he wrote for *Jaws*.

Simon Trezise, in his book *Debussy: La Mer* (Cambridge, 1994) notes that "motifs are constantly propagated by derivation from earlier motifs" (p. 52). Trezise notes that "for much of La Mer, Debussy spurns the more obvious devices associated with the sea, wind, and concomitant storm in favor of his own, highly individual vocabulary" (pp. 48–49). Trezise (p. 53) finds the intrinsic evidence "remarkable." He cautions that no written or reported evidence suggests that Debussy consciously sought such proportions. Caroline Potter (in "Debussy and Nature" in *The Cambridge Companion to Debussy*, p. 149) adds that Debussy's depiction of the sea "avoids monotony by using a multitude of water figurations that could be classified as musical onomatopoeia: they evoke the sensation of swaying movement of waves and suggest the pitter-patter of falling droplets of spray" (and so forth), and—significantly—avoid the arpeggiated triads used by Wagner and Schubert to evoke the movement of water.

We know that musical onomatopoeia is closely associated with specific musical instruments, so Debussy probably chose the instruments with a timbre he felt related to the sea, such as the flute solo at the beginning of the movement at bar 44, which might give the impression of a bird soaring above the sea: "It has a lonely character, possibly a sea bird." Debussy also went a stage further when creating sounds related to the sea.

One of Debussy's greatest attributes is the way he creates musical color. Susan Key, a writer for the LA philharmonic program books, describes how "Debussy achieves his sonorities by augmenting the standard orchestra with some glitter: two harps and a large percussion section. But other musical elements also become agents of color. Harmonic changes serve as color washes; chords dissolve rather than resolve. Short melodic motives rather than fully developed themes sparkle in brief solos, substituting timbre and movement for narrative coherence."

The author, musicologist, and pianist Roy Howatt has observed, in his book *Debussy in Proportion*, that the formal boundaries of La Mer correspond exactly to the mathematical ratios called The Golden Section.

As Paul Henry Lang notes, it's "a vibrating, oscillating, glimmering sound complex, caressing the senses" in which Debussy rarely uses the full mass of the orchestra, but approaches it with delicacy and resourcefulness to "shimmer in a thousand colors." Pierre Boulez calls the result "an infinitely flexible conception of acoustical instrumental relationships" that avoided symmetry, "a development conceived in feelings and irreducible to a formal classical plan."

In 1991, the Japanese composer Toru Takemitsu based a work, "Quotation of Dream: Say Sea, Take Me!" on a theme from La Mer. (Takemitsu said, "I am self-taught, but I consider Debussy my first teacher!")

In 2002, Norwegian composer Geir Jenssen (alias Biosphere) loosely based his ambient album *Shenzhou* around looped samples of La Mer.

Like the sea itself, the listener's imagination is free to illustrate the music, which wanders around, never settling. The composer Erik Satie joked that he liked the part at 11:15 am.

As marine biologist and conservationist Rachel Carson noted, like the sea itself, the surface of Debussy's music hints at the brooding mystery of its depths, and ultimately the profound enigma of life itself—after all, humankind carries the primordial salt of the sea in our blood.

8.10.6 Like a fine wine, La Mer is an extraordinary masterpiece of musical paintings

The entire Debussy La Mer evokes moods and feelings, and isn't meant to follow any kind of story. The three movements have a similar feel, perhaps because some similar building blocks went into them. The first thing heard above the quietly droning basses is a rising progression built on the whole tones, fourths, and fifths, and using the rhythmic figure of a short note on the downbeat moving to a much longer one. Fourths and fifths stacked on each other have a strong, forthright quality (they are the key elements of fanfares) but also a sort of blankness (the open strings of violins, violas, and cellos are tuned in fifths, those of double basses and guitars are in fourths). The fourth and fifths recur throughout the work without calling much attention to them, since they are such a fundamental part of tonal music, but they bring an elemental quality to the music, as if conveying something wide and open and vast—the ocean, for example.

Like a fine wine, La Mer is an extraordinary masterpiece of musical paintings, hanging in space. We have to enjoy them as sensual experiences, thinking both laterally and vertically. The Debussy La Mer is a shimmering musical sketch inspired by the sea. The piece's ethereal vocal quality and impressionistic moods were groundbreaking. Today, La Mer is highly regarded as Debussy's unique employment of these creations has more than certainly gone on to influence many later scores, such as we might find within the film industry because of its suggestive atmosphere. The structure of this piece was quite different from other pieces composed around the same time. Debussy fits his structure around the moods, journey, and life of the sea without any human element, just purely about nature.

We have identified how Debussy has used timbre to suggest color, atmosphere, and emotion in the first movement of La Mer. We have also shown how timbre is employed structurally through the course of the work and how for Debussy, timbre is just as, if not more important than any other parameters in La Mer. La Mer is widely considered one of the greatest orchestral works of the twentieth century because of its powerful creation of colors and Debussy's replication of the ocean. It is a masterpiece of suggestion and subtlety in its rich depiction of the ocean, which combines unusual orchestration with daring impressionistic harmonies. La Mer sounds like nothing before it. The work has proven very influential.

8.10.7 Is there a smile in our voice?

Similar to "timbre," the character or quality of a musical sound or voice is distinct from its pitch and intensity, we should pay attention to vocal quality and vocal expression. Like written in the book titled *What Every Engineer Should Know About Business Communication* for each business presentation, we should ask the following question: Is there a smile in our voice? (see Figure 8.6).

- Does our vocal expression convey enthusiasm, conviction, excitement, anger, joy, or seriousness?
- Is it flat or monotone?
- Do we emphasize key words?
- Is there a smile in our voice?
- Do we vary the tone and pitch of our voice to "underline" the key phrases that we want our audience to remember?

For a free lesson on how to use vocal expression, listen to the anchors on the evening news. Grandma was right. You do only get one chance to make a good first impression. When you're networking, one goal should always be to make the very best first impression you can on the strangers you meet. "Uncommon courtesy" is a key strategy throughout all your networking relationships. Often that relationship begins at a networking or professional association event by interacting with a winning smile.

- A winning, sincere and engaging smile says so much about the person wearing it. That smile clearly announces, "I'm a friendly, helpful and interesting person, someone you'll enjoy meeting."

Vocal expression

Smile is the most influential part of our vocal expression

Smiling is a powerful cure that transmits:
  - Happiness
  - Friendliness
  - Warmth
  - Liking
  - Affiliation

Figure 8.6 Is there a smile in our voice?

- Eye contact is an essential component of your winning smile. Yes, we smile with our eyes as well as our mouth. From 10 feet away, you can make eye contact with a person and reinforce it with your smile.
- During your focused but brief interaction, maintain direct eye contact with that person. Don't scan the room trying to identify who else you might want to meet. Strong eye contact suggests confidence and credibility.
- While you don't smile all the time—it would look insincere or peculiar—you should always have a pleasant, confident, and interested expression. The stress of talking to strangers can easily show up on your face, so try to relax the tension in your facial muscles.
- Look at the person's name tag several times during that initial conversation to reinforce his or her name and employer in your short-term memory.

Are smiles part of your communication toolkit? Under the right circumstances, a smile can speak volumes. For example, Germans will usually smile at strangers (in a shop, for example) to be polite, but don't be offended if they don't—this is just part of a generally reserved culture. A winning smile combined with effective eye contact can get your networking relationship off to a great start and lay the foundation for that positive first impression. Try these techniques out at the next networking event you attend. I'll be looking for you ... and your smile.

8.11 Getting to "Yes"—reaching agreement across the miles

> Question: When is the time in Florida the same as the time in Oregon?
>
> Answer: It is tonight, when western Florida on the Central Standard and eastern Oregon on Mountain Standard Time experience the one hour transition when the Daylight Saving Time ends.

Recommendation for business communication: choosing the right perspective and timing, agreement can be reached across the miles. In order to make our business communication more successful, we need to change our mentality. The key is not only for you to act with openness, honesty, and collaboration, but to motivate and educate others on this approach too. Since our world is flat due to industrial globalization, getting to "Yes"—reaching agreement across the miles—is highly important for effective industrial design engineering.

In their classic text, *Getting to Yes: Negotiating Agreement Without Giving In*, Fisher and Ury describe their four principles for effective negotiation. They also describe three common obstacles to negotiation and discuss ways to overcome those obstacles.

Fisher and Ury explain that a good agreement is one that is wise and efficient, and which improves the parties' relationship. Wise agreements satisfy the parties' interests and are fair and lasting. The authors' goal is to develop a method for reaching good agreements. Negotiations often take the form of positional bargaining. In positional bargaining, each part opens with their position on an issue. The parties then bargain from their separate opening positions to agree on one position. Haggling over a price is a typical example of positional bargaining. Fisher and Ury argue that positional bargaining does not tend to produce good agreements. It is an inefficient means of reaching agreements, and the agreements tend to neglect the parties' interests. It encourages stubbornness and so tends to harm the parties' relationship. Principled negotiation provides a better way of reaching good agreements. Fisher and Ury develop four principles of negotiation. Their process of principled negotiation can be used effectively on almost any type of dispute. Their four principles are:

1. Separate the people from the problem;
2. Focus on interests rather than positions;
3. Generate a variety of options before settling on an agreement; and
4. Insist that the agreement be based on objective criteria. (p. 11)

These principles should be observed at each stage of the negotiation process. The process begins with the analysis of the situation or problem, of the other parties' interests and perceptions, and of the existing options. The next stage is to plan ways of responding to the situation and the other parties. Finally, the parties discuss the problem trying to find a solution on which they can agree.

8.11.1   First principle: Separating people and issues

Fisher and Ury's first principle is to separate the people from the issues. People tend to become personally involved with the issues and with their side's positions. And so they will tend to take responses to those issues and positions as personal attacks. Separating the people from the issues allows the parties to address the issues without damaging their relationship. It also helps them to get a clearer view of the substantive problem. The authors identify three basic sorts of people problems.

- First are differences in perception among the parties. Since most conflicts are based on differing interpretations of the facts, it is crucial for both sides to understand the other's viewpoint. The parties should try to put themselves in the other's place. The parties should not simply assume that their worst fears will become the actions of the other party. Nor should one side blame the other for the problem. Each side should try to make proposals that would be appealing to the other side. The more that the parties are involved in the process, the more likely they are to be involved in and to support the outcome.

- Emotions are a second source of people problems. Negotiation can be a frustrating process. People often react with fear or anger when they feel that their interests are threatened. The first step in dealing with emotions is to acknowledge them, and to try to understand their source. The parties must acknowledge the fact that certain emotions are present, even when they don't see those feelings as reasonable. Dismissing another's feelings as unreasonable is likely to provoke an even more intense emotional response. The parties must allow the other side to express their emotions. They must not react emotionally to emotional outbursts. Symbolic gestures such as apologies or an expression of sympathy can help to defuse strong emotions.

- Communication is the third main source of people problems. Negotiators may not be speaking to each other, but may simply be grandstanding for their respective constituencies. The parties may not be listening to each other, but may instead be planning their own responses. Even when the parties are speaking to each other and are listening, misunderstandings may occur. To combat these problems, the parties should employ active listening. The listeners should give the speaker their full attention, occasionally summarizing the speaker's points to confirm their understanding. It is important to remember that understanding the other's case does not mean agreeing with it. Speakers should direct their speech toward the other parties and keep focused on what they are trying to communicate. Each side should avoid blaming or attacking the other, and should speak about themselves.

Generally the best way to deal with people problems is to prevent them from arising. People problems are less likely to come up if the parties have a good relationship, and think of each other as partners in negotiation rather than as adversaries.

8.11.2   Second principle: Focus on interests

Good agreements focus on the parties' interests, rather than their positions. As Fisher and Ury explain, "Your position is something you have decided upon. Your interests are what caused you to so decide" (p. 42).

Defining a problem in terms of positions means that at least one party will "lose" the dispute. When a problem is defined in terms of the parties' underlying interests, it is often possible to find a solution that satisfies both parties' interests with the following two steps:

- The first step is to identify the parties' interests regarding the issue at hand. This can be done by asking why they hold the positions they do, and by considering why they don't hold some other possible position. Each party usually has a number of different interests underlying their positions. And interests may differ somewhat among the individual members of each side. However, all people will share certain basic interests or needs, such as the need for security and economic well-being.
- Once the parties have identified their interests, they must discuss them together. If a party wants the other side to take their interests into account, that party must explain their interests clearly. The other side will be more motivated to take those interests into account if the first party shows that they are paying attention to the other side's interests. Discussions should look forward to the desired solution, rather than focusing on past events. Parties should keep a clear focus on their interests, but remain open to different proposals and positions.

8.11.3   Third principle: Generate options

Fisher and Ury identify four obstacles to generating creative options for solving a problem. Parties may decide prematurely on an option and so fail to consider alternatives. The parties may be intent on narrowing their options to find the single answer. The parties may define the problem in win-lose terms, assuming that the only options are for one side to win and the other to lose. Or a party may decide that it is up to the other side to come up with a solution to the problem. The authors also suggest the following techniques for overcoming these obstacles and generating creative options:

- First, it is important to separate the invention process from the evaluation stage.
- The parties should come together in an informal atmosphere and brainstorm for all possible solutions to the problem.
- Wild and creative proposals are encouraged. Brainstorming sessions can be made more creative and productive by encouraging the parties to shift between four types of thinking:
  - Stating the problem,
  - Analyzing the problem,

- Considering general approaches, and
- Considering specific actions.
- Parties may suggest partial solutions to the problem.
- Only after a variety of proposals have been made should the group turn to evaluating the ideas. Evaluation should start with the most promising proposals.
- The parties may also refine and improve proposals during evaluation.

Participants can avoid falling into a win-lose mentality by focusing on shared interests. When the parties' interests differ, they should seek options in which those differences can be made compatible or even complementary. The key to reconciling different interests is to "look for items that are of low cost to you and high benefit to them, and vice versa" (p. 79). Each side should try to make proposals that are appealing to the other side, and that the other side would find easy to agree to. To do this it is important to identify the decision makers and target proposals directly toward them. Proposals are easier to agree to when they seem legitimate, or when they are supported by precedent. Threats are usually less effective at motivating agreement than are beneficial offers.

8.11.4   Fourth principle: Use objective criteria

When interests are directly opposed, the parties should use objective criteria to resolve their differences. Allowing such differences to spark a battle of wills will destroy relationships, is inefficient, and is not likely to produce wise agreements. A decision based on reasonable standards makes it easier for the parties to agree and preserve their good relationship.

The first step is to develop objective criteria. Usually, there are a number of different criteria which could be used. The parties must agree which criteria are best for their situation. Criteria should be both legitimate and practical. Scientific findings, professional standards, or legal precedent are possible sources of objective criteria. One way to test for objectivity is to ask if both sides would agree to be bound by those standards. Rather than agreeing on substantive criteria, the parties may create a fair procedure for resolving their dispute. For example, children may fairly divide a piece of cake by having one child cut it, and the other choose their piece. There are three points to keep in mind when using objective criteria:

- First, each issue should be approached as a shared search for objective criteria. Ask for the reasoning behind the other party's suggestions. Using the other parties' reasoning to support your own position can be a powerful way to negotiate.
- Second, each party must keep an open mind. They must be reasonable, and be willing to reconsider their positions when there is reason to.

- Third, while they should be reasonable, negotiators must never give in to pressure, threats, or bribes. When the other party stubbornly refuses to be reasonable, the first party may shift the discussion from a search for substantive criteria to a search for procedural criteria.

8.11.5   Fifth principle: Bottom-line based communication according to Lean manufacturing

According to a CRC Press book titled *Lean Manufacturing: Business Bottom-Line Based*, often negotiators need to establish a "bottom line" in an attempt to protect themselves against a poor agreement. The bottom line is what the party anticipates as the worst acceptable outcome. Negotiators decide in advance of actual negotiations to reject any proposal below that line.

No negotiation method can completely overcome differences in power. However, Fisher and Ury suggest ways to protect the weaker party against a poor agreement, and to help the weaker party make the most of their assets. Fisher and Ury argue against using bottom lines. Because the bottom line figure is decided on in advance of discussions, the figure may be arbitrary or unrealistic. Having already committed oneself to a rigid bottom line also inhibits inventiveness in generating options.

Instead the weaker party should concentrate on assessing their best alternative to a negotiated agreement (BATNA). The authors note that "the reason you negotiate is to produce something better than the results you can obtain without negotiating" (p. 104). The weaker party should reject agreements that would leave them worse off than their BATNA. Without a clear idea of their BATNA, a party is simply negotiating blindly. The BATNA is also key to making the most of existing assets. Power in a negotiation comes from the ability to walk away from negotiations. Thus, the party with the best BATNA is the more powerful party in the negotiation. Generally, the weaker party can take unilateral steps to improve their alternatives to negotiation. They must identify potential opportunities and take steps to further develop those opportunities. The weaker party will have a better understanding of the negotiation context if they also try to estimate the other side's BATNA. Fisher and Ury conclude that "developing your BATNA thus not only enables you to determine what is a minimally acceptable agreement, it will probably raise that minimum" (p. 111).

8.11.6   Sixth principle: Risk engineering and management

The CRC Press book titled *What Every Engineer Should Know About Risk Engineering and Management* can be applied to business negotiation in industrial design engineering.

When the Other Party Won't Use Principled Negotiation

Sometimes the other side refuses to budge from their positions, makes personal attacks, seeks only to maximize their own gains, and generally refuses to partake in principled negotiations. Fisher and Ury describe the following three approaches for dealing with opponents who are stuck in positional bargaining:

- First, one side may simply continue to use the principled approach. The authors point out that this approach is often contagious.
- Second, the principled party may use "negotiation jujutsu" to bring the other party in line. Jujutsu is a Japanese martial art and a method of close combat for defeating an armed and armored opponent in which one uses no weapon or only a short weapon. Jujutsu expresses the philosophy of yielding to an opponent's force rather than trying to oppose force with force. Manipulating an opponent's attack using his force and direction allows jujutsu ka to control the balance of their opponent and hence prevent the opponent from resisting the counterattack. The key is to refuse to respond in kind to their positional bargaining.
  - When the other side attacks, the principle party should not counterattack, but should deflect the attack back onto the problem. Positional bargainers usually attack either by asserting their position or by attacking the other side's ideas or people.
  - When they assert their position, respond by asking for the reasons behind that position.
  - When they attack the other side's ideas, the principle party should take it as constructive criticism and invite further feedback and advice. Personal attacks should be recast as attacks on the problem.

Generally the principled party should use questions and strategic silences to draw the other party out.

However, when the other party still remains stuck in positional bargaining, the one-text approach may be used. In this approach, a third party is brought in. The third party should interview each side separately to determine what their underlying interests are. The third party then assembles a list of their interests and asks each side for their comments and criticisms of the list. She then takes those comments and draws up a proposal. The proposal is given to the parties for comments, redrafted, and returned again for more comments. This process continues until the third party feels that no further improvements can be made. At that point, the parties must decide whether to accept the refined proposal or to abandon negotiations.

When the Other Party Uses Dirty Tricks

Sometimes parties will use unethical or unpleasant tricks in an attempt to gain an advantage in negotiations such as good guy/bad guy routines, uncomfortable seating, and leaks to the media. The best way to respond to such tricky tactics is to explicitly raise the issue in negotiations, and to engage in principled negotiation to establish procedural ground rules for the negotiation.

Fisher and Ury identify the general types of tricky tactics. Parties may engage in deliberate deception about the facts, their authority, or their intentions. The best way to protect against being deceived is to seek verification of the other side's claims. It may help to ask them for further clarification of a claim, or to put the claim in writing. However, in doing this it is very important not to be seen as calling the other party a liar; that is, as making a personal attack. Another common type of tactic is psychological warfare. When the tricky party uses a stressful environment, the principled party should identify the problematic element and suggest a more comfortable or fair change. Subtle personal attacks can be made less effective simply be recognizing them for what they are. Explicitly identifying them to the offending party will often put an end to suck attacks. Threats are a way to apply psychological pressure. The principled negotiator should ignore them where possible, or undertake principled negotiations on the use of threats in the proceedings.

The last class of trick tactics is positional pressure tactics which attempt to structure negotiations so that only one side can make concessions. The tricky side may refuse to negotiate, hoping to use their entry into negotiations as a bargaining chip, or they may open with extreme demands. The principled negotiator should recognize this as a bargaining tactic, and look into their interests in refusing to negotiate. They may escalate their demands for every concession they make. The principled negotiator should explicitly identify this tactic to the participants, and give the parties a chance to consider whether they want to continue negotiations under such conditions. Parties may try to make irrevocable commitments to certain positions, or to make take-it-or-leave-it offers. The principled party may decline to recognize the commitment or the finality of the offer, instead treating them as proposals or expressed interests. Insist that any proposals be evaluated on their merits, and don't hesitate to point out dirty tricks.

chapter nine

On the river of industrial design engineering
Flow of poetic thinking

> We cannot solve our problems with the same thinking we used when we created them.
>
> **Albert Einstein**

9.1 "And quiet flows the don" with integrity

How to perform business communication with integrity when people had access to only very limited information as to what was going on at higher levels, and their perceptions were influenced by their

- Preconceptions,
- Customs,
- Myths, and
- Biases.

"And quiet flows the don" when there was no freedom of press, no Internet, just whatever rumors haphazardly circulated by word of mouth. And the great river still flows with integrity, communicating the people's lives with the artistic power.

For effective communication you must be heard. Someone, or some group, must be listening. No listening, no communication. Optimize listening, and you've got an audience. One of the criteria of attracting attention and being listened to is the character, knowledge, and authority that derive from the integrity of the speaker.

Persuasive advertising provides an example of how this works. It begins by being heard in a crowded room, and advertising's room is crowded in many ways. One technique is to select an authority to be your spokesperson or offer a testimonial. If word of mouth is the best advertising, then this is a close second, providing a better chance to be heard above the others. The lesson applied—how to distinguish you among many—is the one we want to apply.

Figure 9.1 Integrity and Babe Ruth's baseball reliability scorecard. (Courtesy of Library of Congress.)

9.1.1   *Integrity and Babe Ruth's baseball reliability scorecard*

Babe Ruth remains an iconic figure in baseball and in America's culture. His baseball reliability scorecard is very impressive (see Figure 9.1). Long the greatest homerun hitter, he was a giant, nearly literally and most figuratively. Despite his legendary oversized lifestyle, he built integrity as the best in his field. Baseball fans know that even before he became an outfielder so that he could play every day, he was a pitcher, and long-held the World Series record for most wins.

You are a new baseball bat manufacturer. How will you get people to purchase your bat, when the Louisville Slugger is the most well-known name in the game? You want people to listen to you. What do you do? Get The Babe. All he needs to do is pick up your bat, swing that thing around a few times, look at it lovingly, and say, "This is the baseball bat I use when I play for the New York Yankees." Others agreed. In fact,

> Ruth happily accepted many advertising offers. He endorsed everything from cereal to Girl Scout cookies to soap. He had his own line of candy bars and pushed "Babe Ruth" brand All-American, all-cotton underwear though he only wore custom-made silk undershorts. He appeared on advertisements for Old Gold cigarettes despite the fact that he never smoked

anything but cigars. Eventually, Ruth's endorsements became so plentiful that he had to hire a business manager and an accountant in order to keep track of all the money he was making on the side.

The Babe Ruth Times

A Marketable Commodity: Selling Babe Ruth to America

Of course, a paid testimonial may dim the credibility of the message. Even though you have the right person saying the right things, the ultimate effectiveness is reduced because they have been "bought out."

9.1.2 Bully pulpit to communicate and persuade

For the election years, we may recall an oft-used phrase first coined by Theodore Roosevelt: the incumbent president is said to enjoy the benefits of "the bully pulpit." For detractors of any sitting president, the president's message is filtered through likes and dislikes. Yet, with the prestige and authority conferred upon the office, particularly in times of crisis, the president receives due deference; with each word, phrase, and subject magnified in importance. As you rise through your profession, the integrity you bring to the job will provide you with your own bully pulpit upon which to communicate and persuade others.

9.1.3 Where knowledge and authority live

This brings us to integrity, the roof under which knowledge and authority dwell. The greater the perceived integrity of the messenger, the more powerful the message is.

Integrity has multiple meanings, with a thread that forms a connection, such as:

- One is the moral/ethical component. It is the issue of character; of honesty. It is difficult to measure the building blocks that comprise it, and easier to spot the lapses that create damage.
- The other meaning for integrity defines something tangible, and measurable. Integrity defines the solidity of a material, or a structure. A business can be said to have integrity, measured by the profit-and-loss statement, cash flow, and net worth.

The two meanings may converge. A business may have financial integrity, and may have developed intangible assets, called goodwill, based on a reputation of honesty and trust. A history of safety, reliability, and quality or innovation may translate into customer loyalty; all derived from integrity. When that company makes a new product announcement, it will enjoy the essential elements of effective communication—it will have a listening audience.

An example of how integrity provides a person with the ability to communicate by being listened to is illustrated with the product announcements of the late Steve Jobs. It's true; however, that success may confer a transient integrity. A company may be the industry leader one year, and because of the speed of innovation, and the components necessary to maintain competitive advantage, that same company may be operating in the red within several product generations later.

Jim Collins, author of *From Good to Great: Why Some Companies Make the Leap... And Others Don't,* addressed the need for integrity within an organization. Collins gave it a name when he said, "All great companies are brutally honest with themselves." With brutal honesty, it is not merely being honest in what one says, but, "You must confront the most brutal facts of your current reality, whatever they might be." This approach is difficult. We tend to want to look good, at all times. So, Collins's advice may run against what may at first be seen as running counter to conventional wisdom—if not of the marketplace, at least that of the prevailing corporate culture.

Apple, who was aligned with IBM/Motorola's Power PC, made the switch to Intel partly because they saw the advantage to having their computers capable of running software from Microsoft—their competition. It was an admission that relates to brutal honesty. The decision to make change, to say, "I was wrong," "We were wrong," or "We're changing plans," confers a degree of integrity that may translate as an example of leadership, and attract the attention of the right audience: investors, customers, employees, and other relevant stakeholders.

The integrity necessary to express brutal honesty does not necessarily mean the next view is the correct view. But, if listening is essential, it reveals a person or organization who understands the power of listening—and, therefore, of being listened to.

With this integrity, your colleagues, supervisors, board of directors, or wider group of stakeholders or the media will be compelled to listen and evaluate. You have effectively leapt over an essential hurdle in effective business communication, you have been listened to. Even if your view is not the prevailing one, or meets with great push-back, the good news is that you've been listened to. You've been heard. In business, effective communication runs two ways. Feedback to your communications, even if it has the difficult sound of brutal honesty, should be encouraged, and acknowledged. It strengthens one's integrity. And you reap the added benefit of keeping your megaphone for use another day.

9.1.4   Integrity and industrial design engineering

Build your base of knowledge and authority: Become known as the go-to expert in your field. Take classes. Enroll in more training. Document your in-service training. Subscribe to professional journals, blogs, and websites. Write on your subject of expertise. Speak in public. The authority you carry will mean that when you talk, people will listen.

Be a builder-upper, not a tearer-downer: Integrity is a component of leadership. In meetings, be attentive to others, and pay close attention to what they contribute. Credit them for their good ideas. As you grow in your career, take your colleagues with you. Mentor others, and encourage them to follow your model. Allow yourself to extend your expertise in the growth of your department, company, or profession.

The authority you carry will mean that when you talk, people will listen.

Share success, accept failure: Integrity as an aspect of character means that you are generous in success, and capable of shouldering the blame. In the long run, you will be marked as an individual people can trust, and trust is the currency of good business.

The authority you carry will mean that when you talk, people will listen.

9.2   Art of "communicating with a glance": From Impressionism to elevator speech

As shown by Figure 9.2, *Landscape at Chaponval* by Camille Pissarro, painted in 1880, brings a scientific rigor into the conception, composition, and execution of the modern work of art.

As an example of a landscape painting which demonstrates the technique of oil on canvas, this masterpiece of art creates a pictorial language that represents the real world in an abstract way.

- This painting is structured to reflect such a concept: in the real world, we don't see form; we see light.
- The painting is constructed to reveal such an idea: we have to fix our field of vision (FOV), give our FOV a structure, so that our FOV could become art.
- The concept and the idea above provide the basis for Impressionism, a contemporary/modern art which, through painting an image as if it was a glance, would show how an object's surroundings can effect what is seen in a memory.

Impressionism enables us to communicate to an audience in a manner that the audience would see an object if they were to glance at it.

Figure 9.2 *Landscape at Chaponval* by Camille Pissarro, painted in 1880. (Courtesy of www.CamillePissarro.org.)

The art of "communicating with a glance" is similar to an "elevator speech," a very short, persuasive oral presentation—typically about 30 to 60 seconds, or the duration of an elevator ride—designed to introduce a creative idea, product, or service.

An "elevator speech" is a term taken from the early days of the Internet explosion when web development companies needed venture capital. Finance firms were swamped with applications for money and the companies that won the cash were often those with a simple pitch. The best were those that could explain a business proposition to the occupants of an elevator in the time it took them to ride to their floor. In other words, an elevator speech that worked was able to describe and sell an idea in 30 seconds or less. Today, an "elevator speech" can be any kind of short speech that sells an idea, promotes your business or markets you as an individual. Elevator speeches don't happen only in elevators, of course. They can happen at business receptions and engineering conferences, and so on. An elevator speech is typically considered a type of business communication that enables you to make a convincing pitch to persuade potential clients/customers.

An elevator pitch is a brief, persuasive speech that you use to spark interest in what your organization does. You can also use them to create interest in a project, idea, or product—or in yourself. A good elevator pitch should last no longer than a short elevator ride of 20 to 30 seconds, hence the name.

An elevator speech is as essential as a business card. You need to be able to say who you are, what you do, what you are interested in doing, and how you can be a resource to your listeners. If you don't have an elevator speech, people won't know what you really do. They should be interesting, memorable, and succinct. They also need to explain what makes you—or your organization, product, or idea—unique.

9.2.1   When to use an elevator pitch

Some people think that this kind of thing is only useful for salespeople who need to pitch their products and services. But you can also use them in other situations. For example, you can use one to introduce your organization to potential clients or customers. You could use them in your organization to sell a new idea to your CEO, or to tell people about the change initiative that you're leading. You can even craft one to tell people what you do for a living.

9.2.2   Know your audience

Before writing any part of your elevator speech, research your audience. You will be much more likely to succeed if your elevator speech is clearly targeted at the individuals you are speaking to. Having a "generic" elevator pitch is almost certain to fail.

9.2.3   Know yourself

Before you can convince anyone of your proposition you need to know exactly what it is. You need to define precisely what you are offering, what problems you can solve and what benefits you bring to a prospective contact or customers. Answer the following questions:

1. What are your key strengths?
2. What adjectives come to mind to describe you?
3. What is it you are trying to "sell" or let others know about you?
4. Why are you interested in the company or industry the person represents?

9.2.4   Outline your talk

Start an outline of your material using bullet points. You don't need to add any detail at this stage; simply write a few notes to help remind you of what you really want to say. They don't need to be complete sentences. You can use the following questions to start your outline:

1. Who am I?
2. What do I offer?
3. What problem is solved?
4. What are the main contributions I can make?
5. What should the listener do as a result of hearing this?

9.2.5   Create an elevator pitch

It can take some time to get your pitch right. You'll likely go through several versions before finding one that is compelling, and that sounds natural in conversation. Follow these steps to create a great pitch, but bear in mind that you'll need to vary your approach depending on what your pitch is about.

9.2.5.1   Identify your goal

Start by thinking about the objective of your pitch. For instance, do you want to tell potential clients about your organization? Do you have a great new product idea that you want to pitch to an executive? Or do you want a simple and engaging speech to explain what you do for a living?

9.2.5.2   Explain what you do

Start your pitch by describing what your organization does. Focus on the problems that you solve and how you help people. If you can, add information or a statistic that shows the value in what you do.

9.2.6   Finalize your speech

Now that you have your outline of your material, you can finalize the speech. The key to doing this is to expand on the notes you made by writing out each section in full. To help you do this, follow these guidelines:

1. Take each note you made and write a sentence about it.
2. Take each of the sentences and connect them together with additional phrases to make them flow.
3. Go through what you have written and change any long words or jargon into everyday language.
4. Go back through the rewritten material and cut out unnecessary words.
5. Finalize your speech by making sure it is no more than 90 words long.

9.3 *Probabilistic nature of poetic expression*

Our decision-making under uncertainty can be modeled by probabilistic distributions, which facilitate risk communication.

The power of poetic expression is often related to its probabilistic nature, which allows readers' involvement, imagination, and interpretation. This poetic expression is especially influential when presenting the enduring effect of life-changing events. Taking Robert Frost's "Never Again Would Bird's Song Be the Same" as an example,

> He would declare and could himself believe
> That the birds there in all the garden round
> From having heard the daylong voice of Eve
> Had added to their own an oversound,
> Her tone of meaning but without the words.
> Admittedly an eloquence so soft
> Could only have had an influence on birds
> When call or laughter carried it aloft.
> Be that as may be, she was in their song.
> Moreover her voice upon their voices crossed
> Had now persisted in the woods so long
> That probably it would never be lost.
> Never again would birds' song be the same.
> And to do that to birds was why she came.

While subject to many interpretations, the following are often our reflections about the poem:

- The poem refers to Adam, without naming him.
- The poem reveals Eve's sound as if it is a gift to nature.
- Adam's passion for Eve is clearly a starting point for Frost's deliberations.
- Line 9 ("Be that as may be, she was in their song") represents a turning point. Here, Robert Frost starts to inject a new tone, slightly more foreboding.
- By Line 12 ("That probably it would never be lost"), we are meant to hear in "probably" an ironic thrust. We know that, as an enduring consequence, much is soon to be lost.

As shown by the example above, probabilistic poetic expression enables us to communicate the most serious of themes with the simplest language.

9.4   Mid-Autumn night: Would we share the same moon?

9.4.1   Moon and business communication

Once at a friend's home in Switzerland, we recognized that it's the evening of the Mid-Autumn Festival.

With the bright moonlight falling on both of us, the friend told me:

> No matter how far you are from where you came from, you share the same moon with the people there.
> No matter how far you are from where you are going to, you share the same moon with the people there.

I feel that the friend communicated our business relationship with the moon that we share at every Mid-Autumn Festival, as illustrated by the following famous Chinese poem:

Thinking of You

Having been drinking happily over night
I'm drunk
So I write this poem
Remembering my brother, Zi You

When will the moon be clear and bright?
With a cup of wine in my hand, I ask the clear sky.
In the heavens on this night,
I wonder what season it would be?

I'd like to ride the wind to fly home.
Yet I fear the crystal and jade mansions
are much too high and cold for me.
Dancing with my moonlit shadow,
It does not seem like the human world.

The moon rounds the red mansion,
Stoops to silk-pad doors,
Shines upon the sleepless,
Bearing no grudge,
Why does the moon tend to be full when people are apart?

People experience sorrow, joy, separation and reunion,
The moon may be dim or bright, round or crescent shaped,
This imperfection has been going on since the beginning of time.

May we all be blessed with longevity,
Though thousands of miles apart, we are still able to share the
  beauty of the moon together.

Mid-Autumn of the Bing Chen year
Su Shi (January 8, 1037–August 24, 1101)

- Although "the moon may be dim or bright, round or crescent shaped," full of uncertainty.
- Although we often feel puzzled, "Why does the moon tend to be full when people are apart? …This imperfection has been going on since the beginning of time."
- Although we may "wonder what season it would be" due to the seasonal variability of a lunar calendar.
- Although the moon may be "much too high and cold for" us to create environmental sensible products.

It's the moon that engineers a great business communication with the people we wish to connect.

"Though thousands of miles apart, we are still able to share the beauty of the moon together" because we share the same moon at the mid-autumn night.

9.4.2   Case study: What would we learn about cross-cultural communication by staring at the moon?

For every dimension of culture, there are many challenges that could hamper effective communication. When we communicate our culture determines numerous factors including whether we make eye contact or not; the meaning of facial expressions; the meaning of gestures; whether physical contact is acceptable; the level of emotional expression; and our body language. Some cultures such as the Dutch communicate very explicitly (stating clearly and directly what is meant) while others such as British people typically communicate much more implicitly.

While color is often used to communicate meaning, bear in mind that colors have different symbolic meanings in different parts of the world. Here are a few examples:

- In China red symbolizes good luck.
- In India it is the color of purity.
- In most of the Western world it is associated with excitement, danger, love, or passion.

Where white indicates purity and peace in the West, it is associated with funerals in China and Japan.

Different cultures also attribute different meanings to specific symbols and signs, as illustrated in the diverse interpretations of hand signs. Let's look at some examples:

- In South Africa and most parts of the Western world, the good old "thumbs up" indicates agreement or that everything is well; however, in Greece it can also be a very rude "up yours!"
- In Europe this same sign is sometimes used for the number "one", but if you use it in a fruit market in Japan you might get more than you wanted as there it illustrates "five."
- While the "okay" sign has a positive meaning for scuba divers and most people in the United Kingdom and United States, it is considered obscene in Iran, Spain, Greece, parts of Eastern Europe and Latin America. It could also signify "worthless" or "zero" in France.

Even more peculiar is that culture not only has an influence on our interpretation of colors and symbols, but that it even has a strong influence on our visual perception. A practical illustration of this is a study that found that when people look at the dark parts of the moon, there are different visual perceptions:

- Most people in North America looking at the dark parts of the moon see the image of a man; while people in India and China see a rabbit (see Figure 9.3); and Native Americans see a toad.
- In the Southern hemisphere (where we see the moon from a different angle than in the Northern hemisphere) Polynesians see a woman; Fijians see a rat; and Australians see a cat!

Figure 9.3 When looking at the dark parts of the moon, people in India and China see a rabbit.

Of course, on the most basic level, language is probably the first area in which cultural barriers could affect the effectiveness of our cross-cultural communication efforts. Long before the world became so culturally diverse and interactive as it is today, the following four language-related barriers to cross-cultural communication were identified (Munter 1993):

- Semantic barriers: This refers to problems such as people not knowing the meaning of a word, the fact that a word can have different meanings in the same language, and that some words simply cannot be directly translated.
- Connotation barriers: A word can imply different things in different languages. An example is the Japanese word "hai" which is often translated as "yes." However, while this might indicate agreement to an American, the Japanese connotation is closer to "yes, I'm listening," than it is to "yes, I agree." Similarly, while "mañana" in Spanish and "bukara" in Arabic are both translated as "tomorrow" in English, their actual meaning is actually closer to "some time in future."
- Tone barriers: In different cultures different voice tones can imply different meanings. In this way a similar tone can indicate formality or informality, politeness or rebuke, and impersonality or closeness, all depending on the culture. Also, in some cultures, people change their tone depending on where or to whom they're speaking—to superiors at work, to friends in a social context, or to children at home. Not knowing this and using a personal or informal tone in a formal situation can thus be perceived as rude or ill-mannered.
- Perception barriers: These barriers are related to the fact that different cultures perceive the world in different ways. For example, Eskimos have many different words for snow because they can perceive many different types of what the English language simply calls "snow," while Hopi Indians have a different perception of time due to the fact that they don't differentiate between past, present, and future verb tenses.

Now before you become completely despondent and decide to stop speaking to anyone who doesn't know the words of your national anthem—cross-cultural communication is a skill (or set of skills, to be more accurate) that can be learned and improved. In fact, if you're serious about being effective and remaining competitive in this volatile globalized economy, investing in improving these skills might be one of the best strategic moves you've ever made.

9.5   5 Whys for business innovation: Simple yet smart

In business there is only one stupid question! (LinkedIn News, July 5, 2014)
So what is the only stupid question in business?
It's the one you don't ask!
The only stupid question is the one you don't ask!
The author commented:
"… Instead of being the only stupid question in business, would the following be the smartest question in business:

- Why is that so important the idea?
- Why is that so important that idea?

Should we ask each of these Whys 5 times (5 Whys)"

9.5.1   5 whys: A formula to better problem solving

5 whys are used in problem solving. It's a technique of asking a first why. When you find the response, challenge it again with a why. Keep going until you reach level 5. Generally this is the root of your problem. It is commonly named root cause analysis. You can play this but with innovation focus. You can alternate with "Why not?"

Root cause analysis and preventive maintenance are concepts we expect to see in a factory setting. Innovations supposedly don't have time for detailed processes and procedures. And yet the key to startup speed is to maintain a disciplined approach to testing and evaluating new products, features, and ideas. As innovations scale, this agility will be lost unless the founders maintain a consistent investment in that discipline. Techniques from Lean manufacturing can be part of a startup's innovation culture.

As described by the book titled *What Every Engineer Should Know About Business Communication*, one such technique is called the "5 Whys," which has its origins in the Toyota Production System (TPS), and posits that behind every supposedly technical problem is actually a human problem. Taiichi Ohno, the architect of the Toyota Production System in the 1950s, describes the method in his book *Toyota Production System: Beyond Large-Scale Production* as "the basis of Toyota's scientific approach… by repeating why five times, the nature of the problem as well as its solution becomes clear."

Ohno encouraged his team to dig into each problem that arose until they found the root cause. "Observe the production floor without preconceptions," he would advise. "Ask 'why' five times about every matter." There might not be exactly 5 "Why's?" For engineering problem solving, "5" is an arbitrary number to remind you to dig deeper into the problem and get past the surface explanation. A run through the 5 whys analysis, though, is usually enough to get to the heart of the issue.

We come across problems in all sorts of situations in life, but, according to Taiichi Ohno, pioneer of the Toyota Production System in the 1950s, "Having no problems is the biggest problem of all." Ohno saw a problem not as a negative, but, in fact, as "a kaizen (continuous improvement) opportunity in disguise." Whenever one cropped up, he encouraged his staff to explore problems first-hand until the root causes were found. "Observe the production floor without preconceptions," he would advise. "Ask 'why' five times about every matter."

Ohno used the example of a welding robot stopping in the middle of its operation to demonstrate the usefulness of his method, finally arriving at the root cause of the problem through persistent enquiry:

"Why did the robot stop?"

The circuit has overloaded, causing a fuse to blow.

"Why is the circuit overloaded?"

There was insufficient lubrication on the bearings, so they locked up.

"Why was there insufficient lubrication on the bearings?"

The oil pump on the robot is not circulating sufficient oil.

"Why is the pump not circulating sufficient oil?"

The pump intake is clogged with metal shavings.

"Why is the intake clogged with metal shavings?"

There is no filter on the pump.

For industrial design engineering, the "5 Whys" process will help you get to the root of any problem, and make every team member feel understood and included.

Sometimes things don't go according to plan. Tools break, wires get crossed; the best-laid plans fall apart. And on those occasions, it helps to know exactly what happened—so it doesn't happen again. Moments like these are when we turn to a simple but remarkably effective process: The 5 Whys.

It's just as it sounds: A discussion of the unexpected event or challenge that follows one train of thought to its logical conclusion by asking "Why" 5 times to get to the root of the problem. However, it's also a lot deeper than that, too. This is a simple but effective way at getting to deeper insights, underlying issues, and the root of a problem. By asking "Why?" around a given statement—as many times as it makes sense—it enables you to dig below the surface-level assumptions or symptoms of a problem in order to find its root cause. For consumer products, all "why" question probes with consumers ultimately lead there—from socks, to cheese, to cars, to snow blowers, and so on.

9.5.2   Innovation games: Play is the highest form
of industrial design engineering

Innovation is often something we think we have or not. That is not true! Innovation is a skill, and like everything it can be learned and trained. There are different games you can do in order to improve your creativity. It should become a practice of every day.

Play is the highest form of industrial design engineering. The most common game you can play is the simplest one but not the easiest. Let's call it "Link it." In this game, you need to be curious of everything. This is a general mantra in innovation. This game consists by linking things you like together. This allows you to create everything and the craziest think. Imagine you like beer, and you like cats: how can you link it? You see, it's simple but not easy. You must always be aware of what you can link and what you can create.

The other game is "What if…" In this game you can revolutionize the world. What if my coffee machine can be easier to wash? What if instead of touching my phone screen I can talk to it? Let's create Nescafé and Siri. Nescafé is a brand of instant coffee made by Nestlé. Siri is a computer program that works as an intelligent personal assistant and knowledge navigator, part of Apple Inc.'s iOS, watchOS, and tvOS operating systems.

The next game can help to train your ability to watch the world. Train yourself to observe what could be done better, what's not working properly, and find a solution. Let's call this game, "Solve it." There is for sure many things coming to your mind. Think about the last time something drove you crazy—solve it!

Finally, business innovation is often at the intersection of two things/ thinks. Find these intersections, innovate!

9.6   Industrial engineering thinking
versus poetic thinking

Looking at Figure 9.4, a picture photographed by my wife Lisa, I recalled the following classical Chinese poem:

Cold Mountain's Color Tour

Winding up the rocky path in Cold Mountain far;
I see a house amid the white clouds deep.
Stopped my coach to enjoy late autumn's maple woods;
Frosty leaves are redder than spring flowers.

Du Mu (803–852)

Here, the poet uses a joyful tone as he describes the mountain scenery in autumn. He utilizes the maple leaves of late autumn and their fiery red

Figure 9.4 Cold Mountain's color tour (photographed by Lisa Wang).

color to make the mountain forest in late autumn even surpass the spring flowers in full bloom.

Similar to picture thinking in engineers' business communication, picture thinking in poetry introduces abstract thoughts, makes logical or figurative·arguments, or attempts to reach philosophical conclusions via the medium of a picture.

Based on picture thinking, imagism is a movement of poetry that flourished immediately before World War I in the United States and the United Kingdom, the most famous practitioners of which were William Carlos Williams and Amy Lowell. It favored "direct treatment of the thing" in concentrated bursts of imagery and in some ways was modeled on Western ideas of Eastern poetry. In contrast to extraneous descriptions, it attempted to produce a sense of immediacy.

Like a poet, an engineer can think in terms of images, and allow images to produce something like a sequence of ideas or concepts.

Listening to the resonance of "deaf" leaves under the force of "mightiest bear," I read Percy Bysshe Shelley's "Ode to the West Wind," where the process of decay and rebirth has been presented. The "deaf" leaves become seeds that will be reborn, when the spring brings new life to our world.

In risk engineering, such a motif of decay and rebirth can be modeled by renewal theory, which could simulate wear-out including dependence of failures on wear.

...

Comparing with using all components until failure, we can perform scheduled replacement before failure to improve system availability and save cost.

In addition, an engineer can make statements of either an abstract or specific sort, which resemble what we would call "a problem statement" for a problem-solving. Engineers can think logically or analogically (e.g., with mechanical-electrical analogy), using figurative thinking to develop a series of comparisons explicitly or abstractly.

9.7   Hamlet's action: Decision making under uncertainty

Like an engineering project, literature often confronts contradiction and uncertainty. In *The Tragedy of Hamlet, Prince of Denmark*, the action we expect to see from Hamlet himself is continually postponed. Hamlet tries to obtain more certain knowledge about what he is doing although absolute certainty is impossible. Hamlet, a tragic hero, demonstrates his tragic flaw in his indecision.

Directly related to the theme of certainty is the theme of making actionable decision. Is it possible to make purposeful decisions when involving substantial uncertainty?

The book titled *What Every Engineer Should Know About Decision Making Under Uncertainty* presented a Plan–Do–Check–Act (PDCA) model for creating change with significant business impact. Let's illustrate the PDCA model using Hamlet as an example.

- Plan. Prince Hamlet plans to revenge the death of his father (King Hamlet), who was recently murdered. The suspect is Claudius, his uncle, who has inherited the throne and married the king's widow, Queen Gertrude.
- Do. To minimize the uncertainty, Hamlet does an experiment with a play called "The Mousetrap" in order to determine if Claudius is in fact guilty or innocent of the murder. Hamlet tests his uncle's guilt by having a group of traveling actors perform a scene closely resembling the sequence by which Hamlet imagines his uncle (Claudius) to have murdered his father.
- Check. Hamlet and scholar Horatio check Claudius' response to the designed scene. They find that Claudius leaps up and leaves the room. Reviewing the test result, Hamlet and Horatio agree that the experiment proves Claudius' guilt.
- Act. Having confirmed that Claudius is guilty, Hamlet starts to act for revenge.

In Figure 9.5, characters are Hamlet, Horatio, Marcellus, and the ghost of Hamlet's father. The ghost and Marcellus are characterized as follows:

Figure 9.5 Characters are Hamlet, Horatio, Marcellus, and the ghost of Hamlet's father. (Courtesy of Library of Congress.)

- The Ghost: The specter of Hamlet's recently deceased father. The ghost, who claims to have been murdered by Claudius, calls upon Hamlet to avenge him. However, it is not entirely certain whether the ghost is what it appears to be, or whether it is something else. Hamlet speculates that the ghost might be a devil sent to deceive him and tempt him into murder, and the question of what the ghost is or where it comes from is never definitively resolved.
- Marcellus: One of the officers who first sees the ghost walking the ramparts of Elsinore and who summons Horatio to witness it. Marcellus is present when Hamlet first encounters the ghost. Elsinore, called Helsingor in Denmark, is a city in eastern Denmark. It is known for its castle Kronborg, where William Shakespeare's play *Hamlet* is set.

From the perspective of integrated risk management, Hamlet's action is characterized by the following:

9.7.1   Action is contradictory

For Hamlet's action, contradictions exist between

- Thought and deed,
- Reason and intellect, and
- Contemplation and action.

Because he is contemplative and thoughtful by nature, Hamlet delays his action. While Hamlet has the intellectual ability to resolve, his moral integrity impedes him to act. Hamlet struggles with his doubts whether executing Claudius, the current king and his uncle, is the appropriate thing to do. As a consequence, Hamlet's procrastination leads him into a deep melancholy.

Hamlet cannot understand Gertrude's remarriage. Gertrude cannot understand Hamlet's passion for revenge, apparently due to ineffective communication.

9.7.2   Action is arresting

Hamlet needs to select among alternatives, as evidenced by Hamlet's soliloquy "To be or not to be."

Hamlet acknowledges the risk for each alternative.

9.7.3   Various setbacks occur over the course of action

By mistake, Hamlet causes the death of Polonius, the father of Ophelia (his potential wife).

Overwhelmed by her father's death, Ophelia drowns in the river with grief.

Claudius leads Laertes, Polonius's son, to a corrupt form of revenge on Hamlet, a swordfight with his sword-point poisoned.

As a backup plan, Claudius decides to poison a goblet, which he will give Hamlet to drink.

Having been poisoned, like his mother Gertrude, Hamlet dies immediately after achieving his revenge.

9.7.4   Action is decisive

Prior to his death, Hamlet has executed Claudius with the poisoned sword and forces him to drink down the rest of the poisoned wine.

Claudius is doubly punished, appropriately with his own poisons.

Event sequence provides parallels: poisoned king (Hamlet's father), poisoned marriage between Gertrude and Claudius, and poisoned offenders (Laertes and Claudius).

9.7.5   Action results in a triumphant conclusion

Hamlet is carried away in a manner befitting a fallen soldier.

Similar to a masterpiece of literature, a sound engineering project is developed through a rigorous process, echoing Plan, Do, Check, and Act (PDCA) systematically.

9.8   Engineering dialogue on Lake Lucerne: How to think and express our impression poetically

In London's Courtauld Gallery, we can imagine an engineering dialogue on Lake Lucerne, while looking toward Fluelen, Switzerland.

Architects began to explore romantic landscape drawing in Britain and Germany in the 1760s. Caspar David Friedrich described Romantic landscape engineering as "a dialogue with nature," claiming that "the artist should not only paint what he sees before him, but also what he sees in himself." His words encapsulated two central elements of the Romantic conception of landscape:

- Close observation of the natural world, and
- Importance of the imagination.

As shown in Figure 9.6, in the Somerset House, Strand, London, we can find Joseph Mallord William Turner's *On Lake Lucerne, looking towards Fluelen* (ca. 1841?). This watercolor, with scraping out and marks made with the thumb, was painted over graphite on wove paper, 223 × 283 mm.

With a new poetic version, Turner virtually eliminated all the natural details. Instead, Turner painted with broad strokes of blue and grey watercolor, some scraped out, some manipulated by his thumb.

Figure 9.6 Joseph Mallord William Turner's *On Lake Lucerne, looking towards Fluelen*. (Courtesy of The Courtauld Institute of Art.)

How to design more poetically, creatively, and originally? Engineering design is a trade-off between expense and effect. Here, Turner achieved the "poke-yoke" (mistake-proof) effects of the silver moon, rays of moonlight, and shadows on the expanses of water and its mountainous surroundings. For industrial design engineering, poetry is a terrific way to think. It's a valuable exercise to bring to life unexpressed feelings and unexpressed thoughts. The job in poetry is to take language where it's freshest. It is the best kind of history—content at its best.

9.9   Would a poet flip a coin like an engineer?

Poetry Editors & Poets (LinkedIn)

> ... Do you proceed by images or by words when you write?

Discussions related to the following comments by the author, "I often write poems by flipping a coin, which is also called Monte Carlo Simulation…"

When writing a book titled *What Every Engineer Should Know About Decision Making Under Uncertainty*, I used "How Would Engineers Flip Coins" as a section heading. The section is about Monte Carlo Simulation.

As shown in Figure 9.7, Monte Carlo simulation is one of the most common randomized algorithms applied to simulate physical systems. A Monte Carlo simulation is a method of estimating the value of an unknown quantity by making use of the principles of inferential statistics.

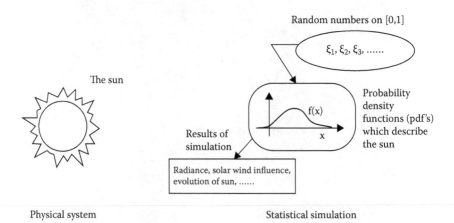

Figure 9.7 Monte Carlo simulation of physical system. (Courtesy of Oak Ridge National Laboratories https://www.phy.ornl.gov/csep/mc/node1.html.)

The guiding principle of inferential statistics is that a random sample tends to exhibit the same properties as the population from which it is drawn. For example, if you flip a coin 10 times, and it comes up heads every single time, then you might start to form an opinion that the coin is not fair, and that it is biased toward coming up heads—you might begin to believe that the population of all possible flips of this coin contained more heads than tails. If you flip the same coin 100 more times, and it still comes up heads every single time, it would probably be hard to convince you that the coin was fair at all. In fact, you might start to believe that it had two heads!

On the other hand, if you flipped the coin 100 times and roughly half of the time it came up heads, then you would probably believe that it was a fair coin—or at least not unfair to any obvious degree.

In both cases, your belief whether the coin is fair is based on the intuition that the behavior of a sample of 100 flips is similar to the behavior of the population of all flips of your coin. Luckily, this belief is not without basis in fact—and we can make use of this fact to achieve good estimates of probabilities with relatively little effort.

The Monte Carlo simulation was conceived in the late 1940s by Stanislaw Ulam, while working on nuclear weapons as a mathematical device for physicists to answer "What if?" questions. Monte Carlo simulation was instrumental in the Manhattan Project, which used the research to develop nuclear weapons during World War II. Ulam's uncle had a fondness for gambling, so since the simulation needed a codename, and they were figuratively spinning the roulette wheel to come up with random numbers, naming it after Monte Carlo seemed appropriate.

Monte Carlo simulation gives industrial design engineers the ability to take that process one stage further by defining the list of feasible values for key inputs and the relative likelihood of each occurrence, and then effectively throwing the dice at least thousands of times to give an idea of the complete range of possible outcomes that would result from these different inputs.

For example, we can run simulation for a hypothetical airline with simplified model inputs for passenger number, average customer spend, fuel price inflation, and other costs. From running a simulator on the best-case scenario and worst-case scenario, we are able to calculate the probability of making a profit for an industrial product.

Probability tells us that the coin will land more or less on heads and tails evenly. If you flip the coin 10 times, it may not be such an equal split, but over such a large scale, we are more likely to get equal results.

Physically flipping a coin 1000 times would be arduous and time consuming. This is where the Monte Carlo simulation will be useful for industrial design engineering, by casting the net over a large amount of data to strengthen the prediction.

9.10   Poem appeared in Poetry Quarterly: Winter/Summer 2012 edition

Written based on my engineering travels to the great rivers of Danube, Aare/Rhine, and Amazon, my poem, "Crossing with Blue Moonlight," appeared in Poetry Quarterly: The Winter/Summer 2012 edition.

Crossing with Blue Moonlight

"Are you drunken
with Sonata Moonlight
on the Beautiful Blue Danube?
Do you remember the blueberry hill?"

I forget
because it's decades ago.
Can you bring me a piece of melody
and a cup of Jasmine tea?

"Are you drunken
with Aare water
under the moonlit dark blue sky?
Do you remember the Blues Trail?"

I forget
because it's oceans away.
Can you bring me a piece of melody
and streams of whiskey?

"Are you drunken
with the turbulent maelstrom
over the river of Amazon?
Do you remember the crossroads
stretching under dark blue sky?
Do you remember the peaceful country roads
crossing with blue moonlight?"

Aare water is the Aare River (Switzerland), which joins Rhine River by the Black Forest. My wife Lisa and I lived by the river when I worked at Paul Scherrer Institute in Switzerland. We also lived in Vienna, Austria, by the beautiful Blue Danube (see Figure 9.8) when I was affiliated with Austrian Aerospace Agency/European Space Agency.

Our global economy relies heavily on creativity for economic growth. A disciplined imagination is fundamental to innovation in all sectors and

Figure 9.8 "Are you drunken with Sonata Moonlight on the Beautiful Blue Danube?"

industries. The market value of services and products derives as much from aesthetic design as from performance.

Poetic thinking provides a competitive advantage across disciplines including industrial design engineering. Organizations seek employees who think imaginatively, solve problems creatively, communicate effectively, and work in teams. Today, a graduate degree in art is very valuable for industrial design engineering. For example, nonverbal forms of knowing exemplified by artists are increasingly important for a company's success in industrial design engineering. Body language, intonation, facial expression, and hand gestures influence business communication. Learning through the arts promotes the unity of objective and subjective knowledge and bridges the gap between self and industrial designs. The arts are complementary with many careers in engineering, engendering conditions cognitive psychologists identify as ideal for learning.

Poetry plays an essential role in the holistic education of industrial design engineers as integrated beings, capable of thinking with and through the mind, heart, and hands. The variety of learning styles promoted by art-making provides the deep cognitive channels required for effectively ordering the self. By projecting ourselves in and through the arts, we design ourselves in salutary ways that align with the human spirit's longing for wholeness. How does this happen?

To study poetry is to learn to "read" images across a spectrum of intellectual, artistic, cultural, and environmental domains, perceiving life through all the mediums in which meaning is expressed:

- Poets could regard their works as a special form of knowledge expressed in a language of images. Ezra Pound made perhaps the most widely used definition of image in the twentieth century: "An 'Image' is that which presents an intellectual and emotional complex in an instant of time." In Pound's definition, the image is not just a stand in for something else; it is a putting-into-words of the emotional, intellectual, and concrete stuff that we experience in any given moment. It is also important to note that an image in poetry, contrary to popular belief, is not simply visual. It can engage any of the senses. And, in fact, for it to be an image, it must engage at least one of the senses by using sensory detail.

- Poets could think in music and not simply about music. What makes a poem sound pleasing to the ear? A solid rhythm for one thing— something we know a good song also depends on. A recognizable song structure in popular music is the African American blues stanza. The intention of the lyrics is to express an emotion. The blues rhythm originated in African American field hollers and work songs, and some trace it to centuries-old songs by griots of West Africa. A griot is a West African historian, storyteller, praise singer, poet, and/or musician. Here is a stanza from Lead Belly's "Good Morning Blues":

Good morning blues. How do you do?
Good morning blues. How do you do?
I'm doing all right. Good morning. How are you?

- Poets could think critically in color and forms, making critical and aesthetic choices, say, with pigment or stone. By thinking in the intersensory domain of images, we connect with our archetypal humanity and the primal impulses behind the aesthetic-artistic experience. For example, *Landscape with the Fall of Icarus* touches on the Greek myth of the tragedy of Icarus, which has been addressed by the book titled *What Every Engineer Should Know About Risk Engineering and Management*. As we know, according to Ovid and Appolodorus, Icarus, son of Daedalus, took flight from imprisonment wearing the fragile wings his father had fashioned for him. Heedless of his father's warning to keep a middle course over the sea and avoid closeness with the sun, the soaring boy exultantly flew too close to the burning sun, which melted his wings so that Icarus hurtled to the sea and death. The death of Icarus, the poet tells us "according to Brueghel," took place in spring when the year was emerging in all its pageantry (see Figure 9.9).

Innovators in industrial designs are "poets" not because they are prone to fantasies or states of dissociation, but because, like children, they harness playfulness to productive ends. Imaginative play is the capacity for shifting the experience of reality. Walt Whitman wrote, "There was a child went forth every day;/And the first object he looked upon, that object he became;/And that object became part of him…." Imaginative play, like art, constitutes a poetic intelligence that is not a matter of right

Figure 9.9 The irony of the death of Icarus, according to Brueghel, is that his death goes unnoticed in the spring—a mere splash in the sea. (Courtesy of Royal Museums of Fine Arts of Belgium.)

answers, the correct use of language, or the accurate reckoning of numbers. Serious play probes the very essence of things, focusing attention on various qualities of being. Cid Corman, an American poet, translator, and editor, highlighted this poetic power of children and artists in all sectors when he advised, "Follow/the stream:/Don't go/but be/going."

Some masters of modern art—Klee, Kandinsky, Picasso, and Miro—found inspiration in the artwork of children. They understood that they must activate their child-like mind, tapping the wellspring of creative consciousness to support their personal and professional well-being. "All children are artists; the trick is to reclaim this when we grow up," stated Picasso. Similarly, Baudelaire knew that "Genius is childhood recovered at will." Children, poets, entrepreneurs, and industrial designers share the principle of creative improvisation, freely using images for purposes of art, work, and life.

Poetic thinking in images allows us to see life as simultaneously literal and imaginable, subject to interpretation and change. Perceiving reality "as if" empowers us to create new worlds with new industrial designs. This "seeing through" the literal and conventional to the possible carries evolutionary thrust, an antidote to the crippling attitude of this is the way things are. Realizing the danger stemming from lack of poetic intelligence, Robert Frost warned that unless we have a "proper poetical education," we are "not safe anywhere." As such, poetic intelligence is a key to unlocking cognitive freedom and an improvisational mind, that is, the capacity to creatively respond effectively to chance and the uncertain availability of resources. As improvisators of life, we combine and recombine familiar and unfamiliar elements of experience in response to shifting situations. Poetic intelligence is vitality, life.

The poems are not simply expressive; they offer and enhance important tools of thinking. For example, image-making is the mind's most fundamental activity of knowing. Learning results when our experiences provide, confirm, or modify images we hold of the world and ourselves. As primary units of consciousness, images exist in all the sensory modes of perception. Images of sight, sound, taste, touch, and smell abound in the mind. Not simply metaphors for ideas, images relate to how we obtain, structure, and use information. This is as true for scientists as it is for artists.

Engineers, like artists, note the function of imagery in their work. A distinctive power of the poetic mind mentioned above is the ability to grasp the inner life of things and not simply to perceive similarities or outer relationships. Eugene Wigner, recipient of the Nobel Prize in physics, said, "The discovery of laws of nature requires, first and foremost, intuition concerning pictures and a great many unconscious processes." The physicist Jerome Friedman, himself a Noble recipient, asserted, "Reasoning is constructed with movable images, just as poetry is." Albert Einstein, who began playing violin at age six, said his discovery of the theory of relativity was the "result of musical perception." Thinking is not done by the brain, but simultaneously in and through the mind-body.

In addition to poetic learning and improvisational thinking through images, prominent poets and engineers—we can add here designers, entrepreneurs, advertisers, and brand marketers—similarly display accurate observation, spatial and kinesthetic thought, identification of core components of a complex whole, recognition of patterns, and empathy with objects of study, as well as visual, verbal, and mathematical synthesis and communication of outcomes. When mathematical, analytical capacities of the mind fuse with aesthetic perception through image-rich arts experience, we undergo holistic transformations that generate inspiring works of animated, embodied thinking and promote our fuller humanity in industrial designs.

9.11   Tulip time, cherry blossom, and risk engineering

While looking forward to the Tulip Time festival in Holland, Michigan, I think of the National Cherry Blossom Festival in Washington, D.C. My wife and I lived in Maryland and the Washington, D.C., area for a few years. Reading the following poem by Anaïs Nin, I feel that the cherry blossom is really like a risk engineering project:

> Risk
> And then the day came,
> when the risk
> to remain tight
> in a bud
> was more painful
> than the risk
> it took
> to Blossom.

Is the cherry blossom a decision-making based on risk engineering?

Two natural facts have had a disproportionate impact on Japanese culture:

- Cherry blossoms are beautiful, and
- They fall.

Cherry blossoms, wrote the scholar-poet Matsudaira Sadanobu (1758–1828), "seem especially suited to the ways of our country, with branches so gentle, flowers so delicate in shape, and hues so simple that the total effect is perfect beyond belief." Sadanobu is famous for his financial reforms by instituting the Kansei reforms, a series of conservative fiscal and social measures.

As an interesting character, Sadanobu influenced Japan's industries. As the shogun's senior counselor from 1787 to 1794 during a particularly acute economic crisis, he "issued an astonishing series of restrictive edicts forbidding almost every form of expenditure by almost every kind of person. ... Ordering women to dress their own hair, he enjoined professional coiffeurs to become washerwomen. ... He awarded prizes for chastity, piety and similar virtues," wrote the eminent British historian George Sansom (Sansom 1978). From the industrial design engineering perspective, Sadanobu was trying to accomplish cost-saving in supply chains.

Our global engineering increasingly relies upon change, innovation, and renewal. A disciplined imagination is a prerequisite for the innovation underpinning value creation in all engineering fields including industrial design engineering. Creativity is the essential ingredient for industrial designs. Today the market value of engineering products and services derives as much from uniqueness and aesthetic appeal as it does from function and performance. Excellent products inspire us with their artistry, design, or beauty.

9.12   The river: Flow of poetic thinking

Do you hate the river
since he died there?
Do you fear the river
since she injured there?

Friends, you really don't understand love
which is simple yet beautiful.
You know, I can always point to the river
and tell you: I was born there.

Poetic thinking examines the human aptitude to approach life and
the worlds we inhabit poetically, that is, in manners of writing, speaking,
art, and industrial design creation that are not governed by systematic
reasoning. Engineering poetic thinking reflects how poetic works across
engineering field raise technical, personal, communal, ethical, and political dilemmas. Through mythological, philosophical, literary, cinematic,
and virtual arts, we can see how poetic works think through or permit
us to deliberate with others about how to live our lives by employing
various rhetorical strategies and creating complex symbolic systems.
Considering how a variety of engineers, designers, thinkers, writers, and
artists approach such notions as "orientation" or "pilgrimage," we will
also examine how they contemplate, poetically, what love, war, death, life,
science, technology, and industrial design itself may mean.

There are two fundamental ways of thinking upon which industrial
design engineering is based: sign-thinking and subject-object-thinking.
These two represent attitudes toward the world which enable us to master
it, at least to a certain degree. However, this power of technology comes at a
high price, since it brings with it distancing and alienation to our ecosystems.
We are not in concordance with ourselves, and we cannot think about the
development of environmental arts and literature and what it means to be
human. In order to integrate this essential aspect of human life into our
conception of the industrial design engineers, we have to develop a poetic
thinking based on thinking language and dialogical thinking. The industrial
design engineers have to develop a wider notion of reason.

For poetic thinking, language is the key to meaning. There is a world
out there but all we can know of it passes through language. Our sensual
perceptions do not mean anything before being integrated in a conceptual
framework—we have to poetize perception into sound and sound into
meaning; there is no direct access to the world. Facts do not exist in isolation in the human world.

Industrial design engineering is based on reality. However, human
beings do not have any direct access to "reality," that is, things (latin: res).

Our reality is always determined by our relation to it. This relation, however, is made up of language. Our perception and sensations have to be translated into world and reality by language, by a discourse that we create over time and into which, at every moment, we have to fit the world we encounter. This is how human beings create their world in dealing mentally with the diffuse perceptions: language forms them in a way that they make sense. The mental dimension is constituted in the process of language—language is not a product, it is always activity—and language is constantly in a mutual relationship with the outer world. A constant interaction between form of life and form of language takes place. The more our use of language widens the limits of our mind and thus of our world, the more it is poetic. Poetic comes from poiesis, etymologically derived from the ancient Greek term ποιέω, which means "to make/ create." This word, the root of our modern "poetry," was first a verb, an action that transforms and continues the world.

Wilhelm von Humboldt differentiated between language as sign and language as language. As sign, language is not considered as a creative process: the sign communicates knowledge, that is, already existing knowledge. In this conception, language serves only as a tool for things outside of language. Considered as language, however, language is a living process, each time anew creating meaning that does not exist independently of this process. It is not words that constitute speech but, on the contrary, the words emerge out of speech, as Humboldt pointed out— only in their continuum they make sense. What happens in language as language is therefore, following Émile Benveniste, not a semiotics where signs are recognized, but a semantics where discourse is understood. Poetic thinking thinks this continuum.

The term "continuum" is central to Henri Meschonnic's theory, building on the original meaning of rhythm to provide a notion with which we can think the organization of the continuum. Rhythm was, before Plato, not the coming and going of the waves of the sea, but it refers rather to the flow of a river, thereby designating a form in the process of change, a non-fixed ephemeral form in movement. Conceiving of speech in those terms manifests the historicity of each moment of speech: it is linked to and determined by its unique situation. If we want to think the specificity of what happens in arts and literature, we need to include this continuum.

This is also relevant for our conception of what it means to be human. It leads to potentialist anthropology since if each moment of speech creates its corresponding reality, it goes beyond the rules and automatisms and hence is the precondition for the development of newness. The term anthropology is here of course used in the sense of philosophical anthropology, based on Kantian pragmatic anthropology, that is the theory of what it means to be human and of the ways human beings "make"

themselves—poiesis again. This historicity of language refers us consequently to infinity and to a potential future at the basis of our processes
of being in the world.

We do have to think language in order to understand the way we exist
in the world, and we do have to think dialogically to understand that our
existence is only conceivable as coexistence. Thinking dialogically is also
linguistically manifest: every language has an interdependent duality
of me and you. While the third person (he/she/it) represents something
else, an object, the second person (you) is part of the sphere of the I. A
human being cannot be conceived of outside of this unity with another
human being—if they are, they are not conceived of as human beings but
as objects, often in biopolitical terms. The I needs a You, being is always
being-with. A human being is not an individual but a dividual; we are
dyadic beings. That is why we are not surrounded by the world but are
with the world, part of it. To overcome the subject-object-thinking with its
distancing, we have to shift our way of thinking away from the concep-
tions of the eye and the hand and toward the ear. What we see and grasp
is always outside of us; what we hear resonates within us. We merge with
sound. We should not simply reason but resonate (exhibit or produce resonance or resonant effects).

A conception of reason that implies the other was developed in the
"dialogical principle" by Buber, who distinguished two modes of being:
the I-It and the I-You. In an I-It-relationship, the other is considered as an
object; in an I-You-relationship, however, both the I and the You emerge
only out of this relationship, while the all-embracing sphere of the 'in-
between' emerges with them. Both modes are necessary but only in the
I-You the human being realizes the world, makes it a world he takes part in
instead of acting on it. Here we are not in a subject-object-relationship anymore: we are dealing with a subject-subject-relationship. The I-You mode
is a moment of presence; that is why it is not verifiable or controllable—we
cannot master it. It always emerges only for a moment; a moment, which
establishes each time a new beginning. It is therefore the 'origin' of life.

Poetic thinking, based on thinking language and dialogical thinking,
thus projects poetological anthropology. The so-called "life sciences" only
touch on discontinuous elements; they cannot think the living continuum.
In order to really think what life is, we need the human sciences which
would have a better claim to make on the term life sciences. But since that
term is now lost, I would suggest that we consider the humanities as "disciplines of meaning." They are in charge of analyzing and directing the
processes through which we make meaning of "facts," and make sense of
them.

As long as the industrial designs function in the logic of the sign—
that is, a word representing something else—and in the logic of a subject-
object relationship, we create dead angles and blind spots in our system

of knowledge. Arts and literature go beyond the sign but they cannot be understood within our conceptual framework of today. Language consists of signs, of course, but there is so much more to language and to life. We have to think beyond the sign. We have to think the continuum.

That is why the industrial design engineering must openly take on poetic thinking to be disciplines of meaning. If they do not stand up for this role, our society will either be meaningless or it will draw meaning from less rationally founded ideologies. Since the former is inconceivable and the latter contains many dangers, the industrial design engineering must change and become disciplines of meaning.

Bibliography

Badiru, A. (Ed.) 2005. *Handbook of Industrial and Systems Engineering*. Boca Raton, FL: CRC Press.

Blanchard, B. S., Fabrycky, W. 2005. *Systems Engineering and Analysis (4th Edition)*. Upper Saddle River, NJ: Prentice-Hall.

Church, E. 2015. Story Behind "Give Me Back My Hometown." http://robynns corner.949starcountry.com/story-behind/february-5-2015-story -behind-give-me-back-my-hometown-by-eric-church/.

Dennis, P. 2002. *Lean Production Simplified: A Plain Language Guide to the World's Most Powerful Production System*. New York: Productivity Press.

Groover, M. 2004. *Fundamentals of Modern Manufacturing*. Hagerstown, MD: Phoenix Color, Inc.

Khan, M. I. 2004. *Industrial Engineering*. New Delhi, India: New Age International (P) Ltd.

Malakooti, B. 2013. *Operations and Production Systems with Multiple Objectives*. New York: John Wiley & Sons.

Munter, M. 1993. Cross-cultural communication for managers. *Business Horizons*, 36 (3), 69–78.

Salvendy, G. (Ed.) 2001. *Handbook of Industrial Engineering: Technology and Operations Management*. New York: Wiley & Sons, Inc.

Sansom, G. B. 1978. *Japan: A Short Cultural History*. Stanford, CA: Stanford University Press. p. 141.

Turner, W., Mize, J. H., Case, K. E., Nazemtz, J. W. 1992. *Introduction to Industrial and Systems Engineering (Third Edition)*. Upper Saddle River, NJ: Prentice Hall.

Wang, J. X. 2002. *What Every Engineer Should Know about Decision Making under Uncertainty*. Boca Raton, FL: CRC Press.

Wang, J. X. 2005. *Engineering Robust Designs with Six Sigma*. Upper Saddle River, NJ: Prentice-Hall.

Wang, J. X. 2008. *What Every Engineer Should Know about Business Communication*. Boca Raton, FL: CRC Press.

Wang, J. X. 2010. *Lean Manufacturing: Business Bottom-Line Based*. Boca Raton, FL: CRC Press.

Wang, J. X. 2012. *Green Electronics Manufacturing: Creating Environmental Sensible Products*. Boca Raton, FL: CRC Press.

Wang, J. X. 2015. *Cellular Manufacturing: Mitigating Risk and Uncertainty*. Boca Raton, FL: CRC Press.

Wang, J. X., Roush, M. L. 2002. *What Every Engineer Should Know about Risk Engineering and Management*. Boca Raton, FL: CRC Press.

Index

Page numbers followed by f and t indicate figures and tables, respectively.